MuPAD

Multi Processing Algebra Data Tool

Tutorial

MuPAD Version 1.2

Benno Fuchssteiner
Klaus Gottheil
Andreas Kemper
Oliver Kluge
Karsten Morisse
Holger Naundorf
Gudrun Oevel
Thorsten Schulze
Waldemar Wiwianka

Birkhäuser Verlag
Basel · Boston · Berlin

Authors' address:
Benno Fuchssteiner
Automath (FB 17)
University of Paderborn
Warburger Strasse 100
D-33095 Paderborn
Germany

A CIP catalogue record for this book is available from the Library of Congress,
Washington D.C., USA

Deutsche Bibliothek Cataloguing-in-Publication Data

Multi processing algebra data tool : MuPAD / Benno
Fuchssteiner ... - Basel ; Boston ; Berlin : Birkhäuser.
NE: Fuchssteiner, Benno; MuPAD

Tutorial : MuPAD Version 1.2. - 1994
  ISBN 3-7643-5017-2 (Basel ...)
  ISBN 0-8176-5017-2 (Boston)

© 1994 Birkhäuser Verlag Basel, P.O. Box 133, CH-4010 Basel
Camera-ready copy prepared by the authors
Printed on acid free paper
Cover design: Markus Etterich, Basel
Printed in Germany
ISBN 3-7643-5017-2
ISBN 0-8176-5017-2
9 8 7 6 5 4 3 2 1

Contents

Chapter 1

Preface

MuPAD is a computer algebra project of the MathPAD group at the University of Paderborn. MuPAD was designed as a parallel system. The design and implementation of MuPAD grew out of the desire to efficiently handle large data generated by algorithms used to investigate the group theoretical structure of nonlinear systems. Nevertheless, MuPAD outgrew this original goal and was developed as a general purpose system and should be used as such.

MuPAD had two major design goals. As already mentioned, firstly we wanted to provide a tool for fast and efficient handling of large data. This goal was motivated by the special problems which came up in our research on nonlinear systems, where data of several GB are not unusual. As a consequence of this MuPAD is a parallel computer algebra system working on the basis of a shared memory machine. Special interfaces, simulating shared memory, will be provided for machines with a different architecture. A sequential version of MuPAD is available which, nevertheless, in its high-end language provides parallel constructs for programming. In this sequential version parallel blocks are executed at random, thus allowing for logical tests of parallel programs on sequential machines. The sequential MuPAD version is the topic of this reference manual.

The second major design goal was to make sure that future versions of MuPAD could be the basis for a system capable of learning during interactive use. Thus, the chosen data structure allows the manipulation of all data, including any programs written in the high-end programming language of MuPAD. The functionality of each MuPAD program and of those functions implemented in the kernel, can be changed according to the need of the user. The advantages and risks of such a strategy will be described in a future publication.

At present MuPAD versions under UNIX and the Macintosh OS have been released. The UNIX version runs either under the usual operating system or with a more comfortable user interface under X (at present we are using the XView Toolkit but we shall change to Motif in the future). In addition we have a prerelease parallel version running on a Sequent Symmetry. This, and other parallel versions, will be released at a future date.

MuPAD consists of basic system functions which, for speed and efficiency, were implemented in the kernel of the system. Further components are a high level language, allowing the user comfortable programming (including parallel constructs) and several interface modules.

The system functions include the necessary arithmetic of arbitrary length. This arithmetic is realized by incorporating the PARI[1] system into MuPAD. A comfortable graphic system which allows efficient computation and display of three-dimensional and two-dimensional functions is included in the system.

Since acceptance of software systems by the user community is often based on comfortable user interfaces, special care has been taken to provide well designed user interfaces which were already distributed with MuPAD's first release. At present the MuPAD interfaces include a comfortable debugger, running either interactively within the system or as a more comfortable separate tool under X. For X, we also provide a comfortable on-line hypertext help system based on the HyTeX[2] system allowing a hierarchical search in this manual and its helpfiles. Furthermore there is the interface for the MuPAD kernel, which is nearly complete under X and less complete in the Macintosh version. This last version, however, will be enhanced in the near future.

This manual describes the functionality of MuPAD from the users point of view only. A technical guide-book describing the design strategy and the technical details of the implementation of the system will follow. The english documentation of MuPAD is divided into two parts, the Reference Manual and this Tutorial.

This tutorial gives an introduction to MuPAD and is addressed to beginners in computer algebra as well as advanced users. A basic knowledge and education in mathematics is required. Experience with other computer algebra packages is useful but not necessary to understand most part of the book.

This tutorial is not meant to be a step-by-step guide to MuPAD, but encourages a user to get MuPAD known through "learning by doing". Therefore, the first part of the tutorial consists of "A sample session", in which fundamental structures and features of MuPAD are introduced. In the second chapter programs written in the MuPAD language are discussed, which give an overview of some specific MuPAD characteristics as well as exhibit possible application areas. The third part of the tutorial summarizes the MuPAD system systematically, whereas the forth chapter concentrates on "Tools and User Interfaces" provided by MuPAD. A list of functionality items completes the tutorial.

MuPAD was developed as a service to the scientific community. As a consequence MuPAD, although copyrighted material[3], has a special distribution policy.

---

[1]PARI, Copyright (c) by C. Batut, D. Bernardi, H. Cohen, and M. Olivier

[2]HyTeX, Copyright (c) by N. Koeckler, University of Paderborn, Germany

[3]MuPAD, Copyright (c) by B. Fuchssteiner, Automath, University of Paderborn, Germany

MuPAD will be distributed to scientific and educational non-profit organizations anywhere in the world free of charge, however a licence has to be acquired. If, in the future, the maintenance and development of MuPAD can not be continued, then the source code will be made public.

Beginning with the distribution of MuPAD version 2.0, research groups working in the area of computer algebra will, under certain restrictions, have access to the source code of the MuPAD system. This also includes the sources of the MuPAD kernel.

Many organisations have supported the development of MuPAD. We are grateful to the Deutsche Forschungsgemeinschaft, the Heinz-Nixdorf-Institute, the Mathematische Forschungsinstitut Oberwolfach, the University of Technology in Loughborough, and finally the University of Paderborn (here especially the University administration). Furthermore we are indebted to those groups who allowed us to incorporate parts of their software in our system, especially the developers of PARI and the HyTeX system. Additionally, we are very grateful to those colleagues in computer algebra who gave us technical and scientific advice, especially to members of the Maple[4] group.

MuPAD is a joint effort of many people, some working full-time on the system and others working marginally on its development. In addition to the authors of this reference manual Birgit Tomann, Gerald Siek, Ralf Hillebrand and Klaus Hering should be mentioned. Mrs. McIntosh-Schneider was of invaluable help in preparing the English manuscript.

In case you would like to obtain a copy of the MuPAD system please send either an e-mail to

MuPAD-distribution@uni-paderborn.de

or a letter to

MuPAD distribution
Automath (FB 17)
University of Paderborn
Warburger Strasse 100
D-33095 Paderborn
Germany

This manual is dedicated to the memory of our friend Waldemar Wiwianka, the project leader of MuPAD from its beginnings, who was tragically killed in an accident in July 1993, at the age of 31. MuPAD owes a great deal to his enthusiastic work and the perspectives he helped to develop.

Paderborn, October 1993

---

[4]MapleV, Copyright (c) 1981-1990 by the University of Waterloo. Maple is a registered trademark of Waterloo Maple Software.

Chapter 2

A sample session

The following examples explain how to work with MuPAD. At the end of this sample session the reader should feel at ease with the system commands and should have some basic understanding of the MuPAD language and the data structures of the system. The sample session is concluded with a collection of tips and tricks, which enable the user to enhance the efficiency of his programs. Many elements of MuPAD are more or less self-explanatory, therefore, no special experience in programming is necessary in order to understand the sample session. Although no knowledge of other computer algebra systems is required, such knowledge will nevertheless facilitate the understanding of MuPAD. Some basic knowledge and education in mathematics is necessary.

During this sample session some auxiliary programs are needed. These programs will be documented in the last section. Their functionality is explained when they are used for the first time. In order to have access to these programs the reader should load them at the beginning of the session with the command `read("tutorial"):`.

Under UNIX MuPAD is started by the command

```
mupad
```

or

```
xmupad
```

However, this depends on the local installation and the chosen hardware. On the Macintosh, MuPAD is started by the usual double-click on the MuPAD icon.

After calling MuPAD the MuPAD logo appears. MuPAD then indicates that it is ready by showing its *Prompt*, which in the usual version has the form >>. This, as well as subsequent prompts, indicates that MuPAD is waiting for the user's input. The screen now has the following form

```
  *----*
  /|   /|      MuPAD 1.2  --  Multi Processing Algebra Data Tool
 *----* |
 | *--|-*      Copyright (c) 1992 by B. Fuchssteiner, Automath
 |/   |/       University of Paderborn.  All rights reserved.
 *----*
```

```
>>
```

Here, as well as in the following, input and output are indicated by using typescript.
Any valid MuPAD statement can now be given as input. The user should start
with the command **read("tutorial")**: followed by <Return>.

2.1   Simple arithmetic

Only the expression to be computed is needed as input in simple arithmetic:

```
>> 2 + 3 * 4;
     14
```

The input 2 + 3 * 4 is entered between the prompt and a semicolon. The output
is then the result 14. Executed input is simplified according to the usual rules
of arithmetic. For example, fractions are simplified by dividing nominator and
denominator by the greatest common divisor:

```
>> 333/450 + 234/900;
      1
```

Input will be executed by pressing either <Return> or <Enter>, depending on
the hardware platform.

MuPAD output can have different forms depending on the variable **PRETTY_PRINT**
which is a so-called *Environment variable*. The notion environment variable means
that by such a variable the user can control and customize his MuPAD environment. If MuPAD output should have the form of valid MuPAD input, the pretty-
printer must be switched off. Otherwise a form of output is chosen which is closer
to the usual mathematical notation. The pretty-printer is switched on and off in
the following way:

```
>> PRETTY_PRINT := FALSE:  (a+b)/(c+d);
     (a+b)/(c+d)
```

```
>> PRETTY_PRINT := TRUE:  (a+b)/(c+d);
```

```
a + b
-------
c + d
```

One can argue a lot about advantages and disadvantages of these two forms of output. If the value of the variable `PRETTY_PRINT` is `TRUE` then, at least for small data, output can be understood more easily. Apart from a few exceptions output with pretty-printer switched off has the advantage that it has the same form as valid input. In this tutorial we use output with the pretty-printer switched off most of the time. For tables and polynomials, we prefer the formatted output obtained with the pretty-printer switched on.

The MuPAD system contains the following arithmetic functions:

Function	Meaning	Example
+	sum	`333 + 42566`
-	difference	`476 - 88384`
*	multiplication	`45*678`
/	division	`45/678`
^	exponentiation	`2^66`
sign()	sign	`sign(-5)`
abs()	absolute value	`abs(-5)`
max()	maximum	`max(4, 5, 6)`
min()	minimum	`min(-2, 6, 78)`
fact()	factorial	`fact(43)`
round()	rounding	`round(3.5)`
ceil()	rounding from above	`ceil(4.3)`
floor()	rounding from below	`floor(4.3)`
trunc()	integer part	`trunc(8/3)`
frac()	fractional part	`frac(8/3)`
float()	conversion to decimal	`float(8/3)`

If all the commands contained in this table are entered, and if the corresponding input is properly ended by a semicolon, the result is:

```
>> 333 + 42566; 476-88384; 45*678; 45/678; 2^66; sign(-5);
abs(-5); max(4, 5, 6); min(-2, 6, 78); fact(43); round(3.5);
ceil(4.3); floor(4.3); trunc(8/3); frac(8/3); float(8/3);
        42899
        - 87908
        30510
        15/226
        73786976294838206464
        - 1
```

```
5
6
- 2
604152630633738356373551320685139975072645120000000000
4
5
4
2
2/3
2.666666666
```

We chose so many commands simultaneously in order to show that they may extend over several lines if the end of line is masked by a backslash (\); however this is not necessary if MuPAD runs under the X-interface or on the Macintosh.

We have seen that MuPAD commands are ended by a semicolon, however, if a command is ended with a colon, the output generated by that command does not appear on the screen. For example, the input of

```
>> 333 + 42566: 476-88384: 45*678: 2^66: sign(-5): abs(-5):
max(4, 5, 6): min(-2, 6, 78): fact(43): ceil(4.3): floor(4.3):
round(3.5): trunc(8/3): frac(8/3): float(8/3):
```

results in no output at all.

Output can be influenced not only by using the environment variable `PRETTY_PRINT` but also by `TEXTWIDTH` controlling the width of in- and output-lines. Output, even when suppressed by use of the colon, is accessible either by %, %n, or `last(n)`. The percentage sign % always accesses the last output whereas %7 or `last(7)` gives access to the seventh from last output. So, after having entered the last input, the use of:

```
>> %; %2; last(4);
```

will result in

```
2.666666666
2.666666666
2/3
```

This may seem surprising at first since we might have expected:

```
2.666666666
2/3
4
```

However, the input of % has lead to a new enumeration of the output. Hence, this phenomenon can be explained by the change in the history of the session.

The user has access to the history of the system by use of `history()`. So if we start a new session, or if we generate the state of a new session by use of the command `reset()` and then enter some commands followed by `history()` the result on the Macintosh will be:

```
>> reset(): round(3.4): fact(3): 2*3: 4 + 8: history();
   1
                                                        12
   2
                                                         6
   3
                                                         6
   4
                                                         3
   5
   6                                        TRUE
```

So the result of **history()** is the output generated by the session up to then. This is independent of whether or not output was suppressed by the use of a colon. Sometimes the command history shows more output than seen in this example. The reason for this is that initialization of MuPAD leads to the automatic execution of a number of commands which are also documented by the history mechanism. The history mechanism of MuPAD does not reach back indefinitely into the past. As a rule only the last twenty output expressions, and in programs only the last 3 expressions, are shown. However, these values can be changed by use of the environment variable **HISTORY**. For example, the command **HISTORY := [25, 12]** tells the history mechanism to record the last 25 interactive output expressions and the last 12 expressions in programs.

Some particularities should be mentioned. Functions such as **max** and **min** only evaluate as far as possible. For example, if a is a non-assigned identifier then the maximum of 1, 2 and a cannot be computed. Nevertheless, no error is returned, the function call is only partially evaluated:

```
>> max(1, 2, a);
   max(a, 2)
```

Here, it can be seen that evaluation was executed as far as possible, a strategy which is also followed by other functions. A completely different result, however, is obtained if a value is assigned to the identifier a:

```
>> a := 14;
   14
```

```
>> max(1, 2, a);
   14
```

The result of conversion into a floating point number by use of **float** depends on the accuracy which is required from the system. Generally ten significant digits are required as default value, however, this value can be changed by use of the environment variable **DIGITS**:

```
>> float(1/3);
   0.3333333333
```

```
>> DIGITS := 50;
   50
```

```
>> float(1/3);
   0.33333333333333333333333333333333333333333333333333
```

```
>> DIGITS:=10:
```

Not all MuPAD functions are independent of each other. Sometimes there are relationships between them. For example, between **trunc** and **frac**:

```
>> frac(8/3) + trunc(8/3);
   8/3
```

The careless use of some MuPAD functions will lead to cumbersome situations. The calling of **fact(10000)** produces output of many pages and blocks the computer for some time. Most of the time is used for producing output on the screen. The actual computation takes much less time. Measuring that with the **time** command:

```
>> time(fact(10000));
   33100
```

yields 33 seconds, i.e., 33100 milliseconds. Of course, this result depends again on the computer platform chosen for running MuPAD.

Calling **fact(10000)** really was a little bit demanding because even 500 factorial is a lengthy number:

```
>> fact(500);
   12201368259911100687012387854230469262535743428031928421924130\
   58838585453731538819976054964475022032818630136164771482035841\
   63337872207817720048078520515932928547790757193933060377296085\
   90862704291745478824249127263443056701732707694610628023104520\
   64421887878946575477714986349943677810376442740338273653974713\
   86477878495438489595537537990423241061271326984327745715546300\
   99772027810145610811883737095310163563244329870295638966289110\
   65897476957208792692887128178007026517450776841071962439039430\
   22536422605234945850129918571501248706961568141625359056693420\
   38130088562492468915641267756544818865065938479517753608940050\
   74523894033579847636394490531306232374906644504882466507594670\
   35862074637925184200459369692981022263971952597190945217823330\
   17569345815085523328207628200234026269078983424517120062077140\
   64097945611612762914595123722991334016955236385094288559201870\
   27433795173014586357570828355780158735432768888680120399882380\
   47021514676054454076635359841744304801289383138968816394874690\
   65881750450692636533817505547812864000000000000000000000000000\
   00000000000000000000000000000000000000000000000000000000000000\
   000000000000000000000000000000000000
```

These examples show that MuPAD, when using rational or natural numbers, can compute with arbitrary length.

2.2 Number theory

In addition to simple arithmetic there are some number theoretic functions and operations of modular arithmetic. These functions comprise of:

Function	Meaning	Example
div	integer division	`9 div 7`
mod	remainder	`9 mod 7`
ifactor	decomposition in prime factors	`ifactor(355355538457)`
igcd	greatest common divisor	`igcd(345, 867)`
igcdex	extended Euclidian algorithm	`igcdex(345, 867)`
ilcm	least common multiple	`ilcm(34, 6)`
ithprime	i-th prime number	`ithprime(56)`
nextprime	next prime number	`nextprime(6778)`
isprime	probabilistic primality test	`isprime(3447)`
phi	number of coprimes	`phi(51)`

Entering them one after the other yields:

```
>> 9 div 7;
    1

>> 9 mod 7;
    2

>> ifactor(355355538452);
    1, 2, 2, 59, 1, 9587, 1, 157061, 1

>> igcd(345, 867);
    3

>> igcdex(345, 867);
    3, - 98, 39

>> ilcm(34, 6);
    102

>> ithprime(56);
    263

>> nextprime(6778);
    6779
```

```
>> isprime(3447);
   FALSE

>> phi(51);
   32
```

Here `div` and `mod` are the usual operations of modular arithmetic. Decomposing a number into its prime factors by use of `ifactor` results in output with the first term representing the sign of the number which was decomposed, and the following terms representing its factors. Each factor is followed by its exponent. Hence from:

```
>> ifactor(355355538452);
   1, 2, 2, 59, 1, 9587, 1, 157061, 1
```

the original argument follows as a product:

```
>> 1*2^2*59^1*9587^1*157061^1;
   355355538452
```

In addition to the greatest common divisor itself, the function `igcdex` gives this as a linear combination of its arguments. The first part of the output is the greatest common divisor followed by the coefficients of the representation as a linear combination of the arguments of `igcdex`, multiplied by the coefficients. So the information contained in:

```
>> igcdex(345, 867);
   3, - 98, 39
```

is the greatest common divisor:

```
>> igcd(345, 867);
   3
```

together with its representation by the original arguments:

```
>> 345*(-98) + 867*(39);
   3
```

The function `ithprime` returns the n-th prime number as a result. Using this the sequence consisting of the first 100 prime numbers is obtained by:

```
>> ithprime(n) $ n=1..100;
   2, 3, 5, 7, 11, 13, 17, 19, 23, 29, 31, 37, 41, 43, 47, 53,
   59, 61, 67, 71, 73, 79, 83, 89, 97, 101, 103, 107, 109, 113,
   127, 131, 137, 139, 149, 151, 157, 163, 167, 173, 179, 181,
   191, 193, 197, 199, 211, 223, 227, 229, 233, 239, 241, 251,
   257, 263, 269, 271, 277, 281, 283, 293, 307, 311, 313, 317,
   331, 337, 347, 349, 353, 359, 367, 373, 379, 383, 389, 397,
   401, 409, 419, 421, 431, 433, 439, 443, 449, 457, 461, 463,
   467, 479, 487, 491, 499, 503, 509, 521, 523, 541
```

Here we used the sequence operator which we shall explain in greater detail later.

The function `nextprime` computes the next prime number following its argument, and `isprime` is a probabilistic primality test. If `isprime` returns **FALSE** then the argument is certainly not a prime number, however, if `isprime` returns **TRUE** then there remains a small probability that its argument has nontrivial factors. The function `phi` is the Euler function returning the number of relatively prime remainder classes for a natural number.

2.3 Roots and transcendental numbers

MuPAD makes all its functions available with arbitrary precision. For example roots, like the square root `sqrt` or the third root:

```
>> sqrt(5);
   5^(1/2)
```

```
>> 5^(1/3);
   5^(1/3)
```

However, these expressions are only numerically evaluated if explicitly desired, either by use of the function `float`:

```
>> float(sqrt(5)); float(5^(1/3));
   2.236067977
   1.709975946
```

or when an argument of these functions is a floating point number:

```
>> sqrt(5.0); 5.0^(1/3); 5^(1/3);
   2.236067977
   1.709975946
   5^(1/3)
```

Apart from roots, MuPAD knows the following transcendental functions:

Function	Meaning	Example
ln	natural logarithm	`ln(2.0)`
exp	exponential function	`exp(5)`
sin	sine	`sin(3.0)`
cos	cosine	`cos(PI)`
tan	tangent	`tan(5.1)`
sinh	hyperbolic sine	`sinh(E)`
cosh	hyperbolic cosine	`cosh(5)`
tanh	hyperbolic tangent	`tanh(3.1)`
asin	inverse sine	`asin(1)`
acos	inverse cosine	`acos(5.0)`

Function	Meaning	Example
atan	inverse tangent	`atan(1/2)`
asinh	inverse hyperbolic sine	`asinh(E)`
acosh	inverse hyperbolic cosine	`acosh(5)`
atanh	inverse hyperbolic tangent	`atanh(3.1)`

On evaluating some of the commands in this table we obtain:

```
>> ln(2.0);
     0.6931471805

>> exp(5);
     exp(5)

>> sin(3.0);
     0.1411200080

>> cos(PI);
     cos(PI)

>> tan(5.1);
     - 2.449389415

>> sinh(E);
     sinh(E)

>> tanh(3.1);
     0.9959493592

>> acos(5.0);
     -2.292431669*I

>> atanh(3.1);
     0.3345248144 + 1.570796326*I
```

When calling `tan(3.1)` our attention is drawn to the output in its typical floating point form `-0.4161665458e-1`. Here `e-1` denotes that the first part of the expression has to be multiplied by 10^{-1}. Therefore the expression `0.1e-10` is equal to 10^{-11}. In analogy to the square root `sqrt`, automatic evaluation only takes place if the argument is a floating point number.

It may be disturbing that `cos(PI)` only returns the function call, although MuPAD knows numbers like `E`, `PI` and `I`:

```
>> float(PI);
     3.141592653
```

```
>> float(E);
     2.718281828
```

```
>> float(I^2);
     - 1.0
```

However, the user can easily change this by using MuPAD's remember mechanism:

```
>> cos(PI) := -1;
     - 1
```

```
>> 3*cos(PI);
     - 3
```

MuPAD will not forget this assignment during the session if the user assigns the value -1 to the function cos. This functionality is already implemented in the shared libraries. It is still unsatisfactory that even after this assignment MuPAD does not know the result of cos(n*PI) for $n \in \mathbb{Z}$.

2.4   Expressions and Identifiers

MuPAD not only works with numerical values but also with identifiers.

```
>> reset():
```

```
>> a + b;
     a+b
```

Here a and b are *identifiers* (data type *DOM_IDENT*), and a + b (data type *DOM_EXPR*) is an *expression* built up by use of these. The MuPAD data type can be checked by use of the system function domtype:

```
>> domtype(a); domtype(b);
     DOM_IDENT
     DOM_IDENT
```

```
>> domtype(a + b);
     DOM_EXPR
```

As already seen, an assignment of values to identifiers and expressions is possible. For example, if a polynomial is assigned to y:

```
>> y := 2 * z + 1;
     z*2+1
```

then a polynomial in y can be assigned to the new variable x:

```
>> x := y^2 - y;
     z*(-2)+(z*2+1)^2-1
```

Obviously, while computing the output, the value for y from the second last assignment was inserted. The result of x therefore is a polynomial in z. It is possible to work in the usual way with these expressions:

```
>> a := b^2 * x + c * y;
     c*(z*2+1)+b^2*(z*(-2)+(z*2+1)^2-1)
```

Here MuPAD has evaluated all variables as far as possible. Also a subsequent assignment to the unassigned identifiers b and c can be carried out. On calling a this subsequent assignment is inserted and the expression is simplified correspondingly:

```
>> b := 2: c := 2: a;
     z*(-4)+(z*2+1)^2*4-2
```

A further expansion, for example the expansion of the brackets, is only achieved by the explicit use of the command **expand**:

```
>> expand(a);
     z*12+z^2*16+2
```

In order to revoke all assignments made in a MuPAD session the system function **reset** has to be used. This command forces MuPAD to resume its starting-up status.

Assignments to single identifiers can be changed by assigning them new values:

```
>> a := 3: a^2;
     9
```

```
>> a := A: a^2;
     A^2
```

Single identifiers can also be reset to their original unassigned state. This is done by assigning them the value NIL:

```
>> a := 1234: a;
     1234
```

```
>> a := NIL: a;
     a
```

Most MuPAD data are expressions, for example, data constructed by algebraic operations or programs written in the MuPAD language. Other, more special data types will be described in later sections.

2.5 Polynomials

In some computer algebra systems polynomials are represented by expressions, for example, the polynomial assigning each number its square is usually represented by the expression `x^2`. Such a representation is possible in MuPAD. However, in order to work more efficiently with polynomials, MuPAD offers a separate data structure for these. The system function `poly` serves to convert expressions into the corresponding data type `DOM_POLY`. So, for example:

```
>> reset():

>> Q:=poly(x^2+3*x*y, [x,y]); P:=poly(x*y^2+3*x*y+3*y, [x,y]);

                    /  2                  \
               poly \ x  + x y 3, [ x, y ] /


                    /    2                        \
               poly \ x y  + x y 3 + y 3, [ x, y ] /
```

generates two polynomials `Q` and `P` with indeterminates `x,y`. The expression corresponding to a polynomial is easily recovered by use of the function `expr`:

```
>> expr(P);
                                 2
                    y 3 + x y 3 + x y
```

With polynomials the user can perform computations in the usual way:

```
>> P+Q;
                /  2     2                      \
           poly \ x  + x y  + x y 6 + y 3, [ x, y ] /

>> P*Q;

      / 3 2     3        2 3      2 2     2         2        \
 poly \x  y  + x  y 3 + x  y  3 + x  y  9 + x  y 3 + x y  9, [x,y]/

>> P^2;

      / 2 4     2 3      2 2        3        2        2        \
  poly \x  y  + x  y  6 + x  y  9 + x y  6 + x y  18 + y  9, [x,y]/

>> diff(Q,x);
                poly( x 2 + y 3, [ x, y ] )
```

A warning is appropriate at this point. MuPAD distinguishes between a polynomial, say `P`, and the corresponding expression `expr(P)` which represents it. Therefore, it also distinguishes between the expression 1 and the polynomial given by this: 

```
>> P+1;
```

$$\text{poly} \left(x\,y^2 + x\,y\,3 + y\,3,\ [\,x,\ y\,] \right) + 1$$

```
>> P+poly(1,[x,y]);
```

$$\text{poly} \left(x\,y^2 + x\,y\,3 + y\,3 + 1,\ [\,x,\ y\,] \right)$$

P+1 is no longer a polynomial, it is now an expression:

```
>> domtype(P+1);
     DOM_EXPR
```

whereas P+poly(1,[x,y]) is again a polynomial:

```
>> domtype(P+poly(1,[x,y]));
     DOM_POLY
```

Moreover, creating a new data type for polynomials has the considerable advantage that the necessary operations are fairly fast, the data are automatically in normal form, and the memory needed for storing large polynomials is considerably less than for the corresponding expressions. In addition, when representing polynomials by expressions, the user will have difficulties with any normal form representation because the internal simplifier automatically sorts sums and products of expressions. Another advantage is that polynomials are automatically given a third operand, namely the coefficient ring over which they are defined. In the case above this is the ring of expressions:

```
>> op(P);
```

$$x\,y^2 + x\,y\,3 + y\,3,\ [\,x,\ y\,],\ \text{Expr}$$

Instead of the ring of expressions the user can choose any other ring. Internally MuPAD knows, for example, the remainder classes for integers, which are represented by the identifiers IntMod(n), where n has to be an integer:

```
>> q:=poly(x^2+12*x*y,[x,y],IntMod(5));
p:=poly(x*y^2+13*x*y+3*y,[x,y],IntMod(5));
```

$$\text{poly} \left(x^2 + x\,y\,2,\ [\,x,\ y\,],\ \text{IntMod}(\,5\,) \right)$$

$$\text{poly} \left(x\,y^2 + x\,y\,(\,-\,2\,) + y\,(\,-\,2\,),\ [\,x,\ y\,],\ \text{IntMod}(\,5\,) \right)$$

Looking at these examples the user can see that, contrary to its modular arithmetic, MuPAD chooses a symmetric representation for remainder classes, i.e. 2 mod 5 is represented by 2 whereas 3 mod 5 is represented by -2.

Of course, for polynomials over rings different from the ring of expressions, MuPAD automatically replaces the arithmetic operations by the corresponding operations in these rings. Therefore adding the polynomial p five times correctly yields the zero polynomial:

```
>> p+p+p+p+p;
      poly( 0, [ x, y ], IntMod( 5 ) )
```

Apart from the rings provided internally by the system the user can use any other self-defined ring with the help of the domain concept treated later in this tutorial.

2.6   The sequence operator $

In the last example of the number theory section 2.2 the sequence operator $ was used. The command for generating an expression sequence consists of an expression (possibly dependent on an index), the sequence operator $ and the range of the index. The result is the sequence of the corresponding expressions dependent on the index. For example, the first ten square numbers are obtained by:

```
>> i^2 $ i=1..10;
      1, 4, 9, 16, 25, 36, 49, 64, 81, 100
```

and the first seven random numbers are generated by MuPAD's random number generator **random**:

```
>> random() $ i=1..7;
845473509473, 676470788342, 281338779191, 792495900393,
751209539300, 628363443011, 313746086538
```

In the last example we saw that it is not necessary for the expression to really depend on an index. The old-fashioned teacher's special assignment for the pupil to write "I am not supposed to contradict my teacher" six times is carried out by the following command:

```
>> print("I am not supposed to contradict my teacher") $ i=1..6;
      "I am not supposed to contradict my teacher"
      "I am not supposed to contradict my teacher"
      "I am not supposed to contradict my teacher"
      "I am not supposed to contradict my teacher"
      "I am not supposed to contradict my teacher"
      "I am not supposed to contradict my teacher"
```

The sequence operator can also be used to write small routines. So, if MuPAD would not have a built-in factorial function, then **126!** is easily computed by use of the sequence operator:

```
>> a := 1: (a := a*(i + 1)) $ i=1..125: a;
     23721732428800468856771473051394170805702085973808045661837\
     77170052497697783313457227249544076486314839447086187187275\
     19400401837013955325179315652376928996065123321190898603130\
     80000000000000000000000000000000000
```

Here we initialized **a** with **a := 1:** and then generated a sequence of assignments, where we changed **a** in each step by multiplication with the corresponding factor. Finally, we asked MuPAD to give the last value of **a** as output. When working with assignments in the sequence generator, we should not forget to put assignments in brackets, otherwise they are not considered to be algebraic expressions.

The same task, computing **126!**, can be achieved by use of the **for**-loop:

```
>> a := 1: for i from 1 to 125 do a := a*(i + 1) end_for: a;
     23721732428800468856771473051394170805702085973808045661837\
     77170052497697783313457227249544076486314839447086187187275\
     19400401837013955325179315652376928996065123321190898603130\
     80000000000000000000000000000000000
```

At this point we are not too far away from writing real programs. It should be mentioned that the factorial program we have just written is not a very fast one. For the computation of the 1650-digit number **701!** it needed 3 seconds whereas the corresponding system function needed only 116 milliseconds for the same computation:

```
>> time(fact(701));
     116
```

Of course, these times depend on the hardware platform which was chosen for running MuPAD. Later, we shall demonstrate that, for computing factorials, even programs written in the MuPAD language can be speeded up considerably by using either the so-called underline functions or the evaluation mechanism in a more sophisticated form.

While using the sequence operator the user should keep in mind that the index must be an unassigned identifier, otherwise an error occurs:

```
>> i := 2: i^2 $ i=1..5;
     Error: Illegal parameter [_seqgen]
```

This behavior is not due to negligence in the design of the system, but is intentional. For example, if a sequence produces an error then it is desirable to find out which sequence element generated the errror. Therefore this functionality is necessary, in order to obtain information about the value of the running index after a computation has been stopped. If the user wants to work with assigned indices, then a `hold` must be used:

```
>> i := 2: i^2 $ hold(i)=1..5;
      1, 4, 9, 16, 25
```

2.7   Data types

Apart from data types like identifiers and expressions we have already seen some numerical data types. The data type of a MuPAD datum can be determined by use of the function `domtype`:

```
>> domtype(2.3);
      DOM_FLOAT
```

```
>> domtype(3/2);
      DOM_RAT
```

```
>> domtype(7);
      DOM_INT
```

In many cases operations between data are permitted even if they are of a different data type. This is possible, for example, between numerical data of different types:

```
>> 3/2 + 4.2*I;
      3/2 + 4.200000000*I
```

Here the result is a complex number:

```
>> domtype(%);
      DOM_COMPLEX
```

Operations between numbers and identifiers are also possible:

```
>> d+4;
      d+4
```

```
>> domtype(%);
      DOM_EXPR
```

Here the result is an expression.

Of course, it is necessary to be able to link numbers and expressions by arithmetical operations. This is important because in a later assignment these expressions may be evaluated as numbers. Where such operations between different data types do not make sense, they are not possible:

```
>> 4+"f";
      Error: Incompatible operands
```

Here, by using inverted commas `"f"` is declared a string:

```
>> domtype("f");
      DOM_STRING
```

Hence the operation `4+"f"` never can make sense, so an error is reported.

MuPAD knows about lists and is able to work with them:

```
>> reset():
```

```
>> L := [a, b];
      [a, b]
```

```
>> domtype(%);
      DOM_LIST
```

When working with lists the user should always keep in mind that so-called *flattening* takes place. This can be seen in the following example:

```
>> L := [e, b]: e := (c, d);
      c, d
```

```
>> L;
      [c, d, b]
```

Lists may be combined by use of the concatenation operator which is represented by a dot:

```
>> [a, g].[f, k];
      [a, g, f, k]
```

Appending elements to a list is carried out with the system function **append**:

```
>> append([a, b], c);
      [a, b, c]
```

Apart from that, there are further data types which easily allow the implementation of mathematical structures. For example, there are sets:

```
>> {a, b, c};
      {a, b, c}
```

```
>> domtype(%);
      DOM_SET
```

as well as arrays:

```
>> A := array(1..2, 1..2, (1,2)=123);
   array( 1 .. 2, 1 .. 2, (1,2) = 123)
```

```
>> domtype(%);
   DOM_ARRAY
```

According to their mathematical meaning there are special rules for some data structures. For example, multiple entries of the same elements disappear in sets:

```
>> {1, 2, 3, 2};
   {1, 2, 3}
```

Since the order of elements within a set is irrelevant, sets which differ only in the order of their elements are recognized as being equal:

```
>> bool({C, B}={B, C});
   TRUE
```

Of course, sets may contain elements other than identifiers and numbers, for example lists and arrays. In fact, all MuPAD data may be used as elements in sets. This is seen in the array A defined above:

```
>> {A, "string", {j, K}};
   {array(1 .. 2, 1 .. 2, (1,2) = 123), {K, j}, "string"}
```

MuPAD knows the usual operations for sets:

```
>> reset();
```

```
>> M1 := {a, b, c, d}: M2 := {A, B, c, d}:
```

```
>> M1 union M2;
   {a, b, c, d, A, D}
```

```
>> M1 minus M2;
   {a, b}
```

In addition to flattening in lists, flattening of entries takes alsoalso place in expression sequences and sets:

```
>> ((1, 2), g);
   1, 2, g
```

```
>> a := (1, 2): M := {a, b}: M;
   {b, 1, 2}
```

In lists, as well as in expression sequences and arrays, the elements may be addressed individually:

```
>> A := [1, 4, 55664]:
```

```
>> B := array(1..2, 1..3, 1..5, (1,2,4)=423, (2,1,1)=TTT):

>> a := (1, 2, 3):

>> A[3];
      55664

>> B[1, 2, 4];
      423

>> a[2];
      2
```

This does not make sense for sets, since the order of the elements is irrelevant:

```
>> M := {1, 2}:  M[1];
      Error: unknown error [_index]
```

In the array B defined above, the user should observe that arrays need not necessarily have two dimensions but may be of any dimension. Furthermore, not all entries need be defined. Accessing an undefined entry returns a datum of the type *DOM_EXPR*:

```
>> B[1, 2, 3];
      B[1, 2, 3]

>> domtype(%);
      DOM_EXPR
```

whereas calling on a nonexistent element of a list returns an error:

```
>> a := NIL:

>> L := [a, b, c]: L[4];
      Error: Invalid index [list]
```

A table is a very versatile data type. In tables, a value can be assigned to any kind of index, i.e. all MuPAD data can be used as indices for tables. Tables are created automatically by assigning a value to an index. However, this only occurs if the corresponding identifier is not defined as a table, an array, or a list:

```
>> T[1] := value: T;
      table(
            1 = value
            )
```

Extensions of tables are carried out by giving additional assignments to indices:

```
>> T[MUPAD] := fun: T;
   table(
        MUPAD = fun
        1 = value,
        )
```

Entries of tables are accessed by their indices:

```
>> T[MUPAD];
   fun
```

If the index of a table is changed by a later assignment:

```
>> Tab[s] := a: s := 34: Tab[34]; Tab[s];
   Tab[34]
   Tab[34]
```

then, at first, it may seem as if the original assignment has been lost. However, since the table still shows the old entry:

```
>> Tab;
   table(
        s = a
        )
```

there must be a way to access it. Indeed, this is simply done by **hold** which prevents the evaluation of the index:

```
>> Tab[hold(s)];
   a
```

Some further data types will be discussed in greater detail later, here only the most important ones will be treated.

Functions and procedures always return well defined data types. Even empty output has its type, namely *DOM_NULL*:

```
>> domtype(print());
   DOM_NULL
```

```
>> domtype(null());
   DOM_NULL
```

The user should not confuse this with data like **NIL** which are of data type *DOM_NIL*. This type is used to reset identifiers. The type *DOM_NULL* can also be generated by use of the function **null**, which has the simple function of not doing anything.

The data type most often used in a computer algebra system is that of an expression. This type, therefore, is divided into a number of subtypes. These can be determined by use of the function **type**:

```
>> domtype(a*b); domtype(a+b);
     DOM_EXPR
     DOM_EXPR

>> type(a*b);
    "_mult"

>> type(a+b);
    "_plus"

>> type(a^b);
    "_power"
```

This function reveals the construction with which construction the expression has been created. The subdivision given by `type` is rather detailed. There are over 50 different expression types. Every system function can be the result of a `type` call. For instance `"_plus"`, `"sin"` or `"abs"`.

2.8   Domains

When designing algorithms the user often wants to introduce new data types either to extend functionality or because he wants them to behave differently under already defined operations.

One way to realize this is to define a new data type inside the MuPAD kernel. However, normal users are neither willing nor able to modify the kernel, nor is that part of MuPAD generally accessible to them. Therefore MuPAD offers the possibility to extend its data structure on the level of the MuPAD language. This enables object oriented programming on the level of the MuPAD language with newly defined domains. Thus, the user can write his algorithms in polymorphic style.

A domain is a new data type which defines how its elements behave with respect to those functions either already available in the MuPAD kernel or defined by the user. For example, when defining a new domain the user can specify how addition and multiplication of its elements shall work or what effect the exponential function shall have on these new data.

Furthermore, the user can specify how domains are evaluated, what their output shall look like, and so on.

Instead of giving an abstract introduction we can illustrate MuPAD's far reaching domain concept with one basic example. More are to be found in the next chapter.

We start by defining the ring of remainder classes modulo 7. Here, obviously, we not only have to define these classes but we also have to specify how these classes have to behave under the arithmetic operations. Hence with the definition of these new data objects a modification of operations implemented in the MuPAD kernel is necessary, here the modification of operations like $\{+, *, -\}$ and so on.

In order to define a new domain MuPAD provides the functions **domain** and **domattr**. The function **domattr** specifies new attributes for domains. This function can be invoked as an operator `::`, as seen in the following. By the way, **domattr** is not only useful for defining new domain attributes, but can also be used for reading attributes.

We start by using the function **domain** for the definition of the new ring Z7 of remainder classes modulo 7:

```
>> Z7 := domain():
```

This domain needs a name which is stored under its attribute **name**:

```
>> Z7::name := "Z7":
```

In order to generate new elements of this domain we have to specify a suitable method under its attribute **new**:

```
>> Z7::new := proc(x)
          begin
             new(Z7, x mod 7);
          end_proc:
```

However, if now a new element, say **newel**, is defined as the remainder class of **12**:

```
>> newel:= Z7::new(12);
     new(Z7, 5)
```

then the output is difficult to understand. Therefore, we need to change the appearance of such a domain element on screen. This is done by defining the attribute **print** of that domain:

```
>> Z7::print := proc(x)
                local expr;
                begin
                   expr := subs(hold(Z7(q)), hold(q)=extop(x,1));
                   print(expr);
                end_proc:
```

Now the output has a more compact form:

```
>> newel;
     Z7(5)
```

Here the system function **extop** was used which gives access to a domain element. In general, a domain element, say with the name **domel**, can be understood as a sequence of operands which contains all the information necessary for its definition. The call **extop(domel)** returns the operands of this sequence. If the system function op is not overloaded, it also gives access to these operands. Each domain element has a zero operand, namely the domain to which that element belongs. For example, if an element of Z7 is defined as:

```
>> domel := Z7::new(2);
      Z7(2)
```

then extop returns its parts in the following way:

```
>> extop(domel,0);extop(domel,1);
      Z7
      2
```

In order to ensure the usual operators for addition and multiplication are available, we have to overload the corresponding system functions:

```
>> Z7::_plus := proc()
            local x, i;
            begin
               x := extop(args(1), 1);
               for i from 2 to args(0) do
                  x := (x + extop(args(i), 1)) mod 7;
               end_for;
               Z7::new(x);
            end_proc:

Z7::_mult := proc()
            local x, i;
            begin
               x := extop(args(1), 1);
               for i from 2 to args(0) do
                  x := (x * extop(args(i), 1)) mod 7;
               end_for;
               Z7::new(x);
            end_proc:
```

The same has to be done for the functions which test and recognize data types:

```
>> Z7::type := proc(x)
            begin
               "Z7"
            end_proc:

Z7::testtype := proc(x,y)
               begin
                  bool(extop(x,0) = y);
               end_proc:
```

```
Z7::domtype := proc(x)
                begin
                  Z7
                end_proc:
```

Now the new domain has been established and the first computations are possible. In order to do this we must first define some additional elements of the new domain:

```
>> a2 := Z7::new(2); a3 := Z7::new(3);
a4 := Z7::new(4); a5 := Z7::new(5);
     Z7(2)
     Z7(3)
     Z7(4)
     Z7(5)
```

Addition and multiplication for these elements can then be written in the usual way:

```
>> a2 + a4; a2 + a3 + a4 + a5;
     Z7(6)
     Z7(0)
```

```
>> a4 * a5; a2 * a3 * a4 * a5;
     Z7(6)
     Z7(1)
```

Certainly, using the usual notation for arithmetic operations in this new domain is very convenient. However, from the viewpoint of run-time, the efficiency of these operations can be increased by using the direct call of the procedures defined for these domains. So instead of `a1 + a2` the call `Z7::_plus(a1, a2)` increases the speed somewhat.

The reason for this effect is to be seen in the way operations work on domains: First the system function `_plus` checks its arguments, then, on discovering that it has to do with an element of Z7, it evokes `Z7::_plus` with the given arguments. When calling `Z7::_plus` directly, no check of the arguments is necessary and the processing speed is increased.

Domains are suitable for implementing different algebraic categories, for example rings or fields. If in another category procedure the notion *field* has already been defined, then in order to consider Z7 as an element of that category *unit* and *one* elements `Z7::zero` and `Z7::one` have to be defined. This of course is done in the same way:

```
>> Z7::zero := Z7::new(0):
Z7::one := Z7::new(1):
```

We proceed similarly with any other necessary operations, for example with division, inverse w.r.t. addition and multiplication and exponentiation:

```
>> Z7::divide := proc(x,y)
              begin
                Z7::new((extop(x,1) / extop(x,2)) mod 7);
              end_proc:

Z7::negate := proc(x)
              begin
                Z7::new( - extop(x,1) mod 7)
              end_proc:

Z7::invert := proc(x)
              begin
                Z7::new((1/extop(x,1)) mod 7)
              end_proc:

Z7::_power := proc(x,y)
              local help, i;
              begin
                if y=0 then return (Z7::one) end_if;
                if y>0 then
                   help := x;
                else
                   help := Z7::invert(x);
                   x := help;
                end_if;
                for i from 2 to abs(y) do
                   help := Z7::_mult(help, x);
                end_for;
              end_proc:

>> a2^2 + a3^3 - a4;
     6
```

Those who dislike the notation we have chosen above for the elements of Z7,
can, of course, also represent these elements by their remainder classes given by
the numbers from 0 to 6. For this the user only has to change the procedure
Z7::print:

```
>> Z7::print := proc(x)
              local expr;
              begin
                extop(x,1)
              end_proc:
```

If now two elements are defined

```
>> a := Z7::new(34);
b := Z7::new(411);
     6
     5
```

then we obtain correctly:

```
>> a+b;
     4
```

Further examples and particularities of domains can be found in chapter 3.

2.9 Data Structures

As is the case with other computer algebra systems, in MuPAD data are represented by trees. A reasonable knowledge of the data structure is a prerequisite for working efficiently with the system.

For example, internally the algebraic expression a * b + b * d ^ 2 * e has the form:

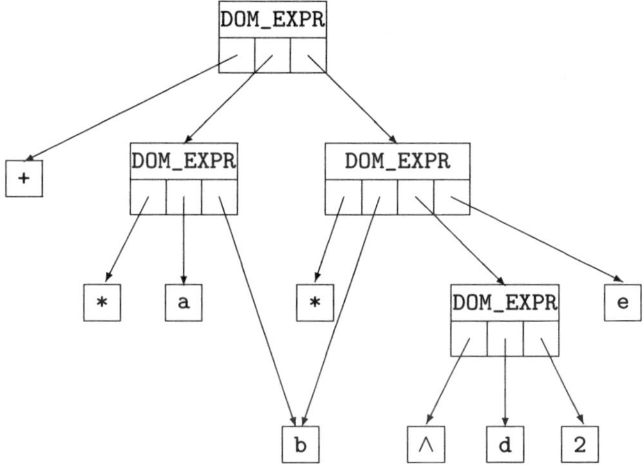

Figure 2.1: Algebraic expression a * b + b * d ^ 2 * e

Here the divided boxes represent internal nodes of the corresponding tree. The division of each box indicates that there is further information hidden which is not accessible to the user. The boxes, together with their entries, are the *leaves* of the tree. In nature, of course, no two nodes can point to the same leaf; however, due to the principle of unique data representation, this is different in MuPAD. If the same datum appears in several places then, very often, it exists physically only once. This strategy saves memory and computation time.

The leaves, as well as the subtrees, can be accessed by the system function op. For this the path leading to the corresponding leave or subtree has to be specified as the second argument in op. Leaves, as well as a subtrees, which start at a node are called *operands*. The path of an operand is represented by a list of positive integers, with each integer representing the number of the path which has to be chosen at the corresponding node. The paths are counted from left to right. The numbering of paths generally starts with 1. However, if the data type is *DOM_EXPR* or *DOM_ARRAY* then it starts with 0.

In the example above, the identifier e was reached by the path [2, 3]:

```
>> reset():
```

```
>> op(a * b + b * e* d^2 , [2, 3]);
      e
```

The same leaf could have also been reached by a number of subsequent calls to op so that these calls only have paths consisting of single integers:

```
>> op(op(a * b + b * e * d^2, [2]), [3]);
      e
```

It should be mentioned, that sometimes operands are numbered in a different way than given by the user:

```
>> op(a*b+y, [2]);
      a*b
```

The reason for this is, that the argument of the op function is transformed into a normal form before the call is executed. In this case the normal form of the expression is:

```
>> a*b+y;
      y+a*b
```

The whole tree in figure 2.1 is of data type *DOM_EXPR*, and therefore must have a zero operand. This zero operand is easily found by use of op:

```
>> op(a * b + b * d^2 * e, [0]);
      _plus
```

The result is the function name of the operation used at the highest level to form the expression. For example, if an expression is formed by application of the sine function then the name of that function is returned as the zero operand:

```
>> op(sin(a*b), 0);
      sin
```

Here we can see that if the path consists of only one number then it need not be entered in square brackets. The general rule for a zero operand is: if an expression is formed at its highest level by an operation or function, then the name of that function or operation appears as the zero operand. Operations like + or * also have function names which can be found by use of **op**. These are the so-called *underline functions*. The name of the operation * is found by analysis of the expression a * b + b * d ^ 2 * e by taking either the path [2, 0] or [1, 0]:

```
>> op(a * b + b * d^2 * e, [1, 0]);
    _mult

>> op(a * b + b * d^2 * e, [2, 0]);
    _mult
```

For arrays, the zero operand is an expression sequence consisting of the dimension and index range of that array:

```
>> Matrix := array(1..2, 1..3, 1..7, (2,3,3) = 9):

>> op(Matrix, 0);
    3, 1..2, 1..3, 1..7
```

Whenever a whole subtree is to be obtained, for example, the subtree being accessible by the second path from above, then only the corresponding path number need be given:

```
>> op(a * b + b * d^2 * e, [2]);
    b*d^2*e
```

It has already been mentioned that if the path consists of only one number then the square brackets can be left out:

```
>> op(a * b + b * d^2 * e, 2);
    b*d^2*e
```

The user has to be careful because not all trees have a zero operand, for example data of the type *DOM_LIST*:

```
>> op([a, b], 0);
    FAIL
```

Depending on the environment variable **ERRORLEVEL**, the functions **op** and **subsop** return either **FAIL** or an error message. The number of paths starting at the highest node, i.e., the number of operands, can be found by use of the system function **nops**:

```
>> nops(a * b + b * d^2 * e);
    2
```

However, the user should note that the zero operand, because of its special role, is not counted. To find the number of paths starting at an arbitrary node the user has to combine the functions **op** and **nops**. For example, in the case of the second node at the second highest level of the expression represented in figure 2.1:

```
>> nops(op(a * b + b * d^2 * e, 2));
      3
```

If the sequence of all the operands of an expression is desired then the function op should be used without giving any path:

```
>> op(a*b*c*d*e);
      a, b, c, d, e
```

This is a powerful tool for efficient programming. For example, to convert a sum into a product the user can proceed as follows,

```
>> a1 + a2 + a3 + a4 + a5 + a6 + a7 + a8 + a9;
      a1+a2+a3+a4+a5+a6+a7+a8+a9

>> _mult(op(%));
      a1*a2*a3*a4*a5*a6*a7*a8*a9
```

whereby only the internal function name of the operation * has to be used.

There is a good reason for not counting the zero operands in **nops**. However, if a different behavior is desired, a new procedure which does exactly that is easily written. For example, the procedure **nops_own** which can be found in the file `"tutorial"`. If there is a zero operand then this procedure returns a value increased by 1 compared to that of **nops**:

```
>> nops_own(a * b + b * d^2 * e);
      3

>> nops(a * b + b * d^2 * e);
      2
```

In all other cases its values coincide with those of **nops**:

```
>> nops_own(a);
      1

>> nops(a);
      1
```

The user has to know the structure of MuPAD data in order to execute substitutions correctly. For replacing a complete subtree, there is the command **subs**:

```
>> ausdr := A^G*(A+h);
      A^G*(A+h)

>> subs(ausdr, A^G = 123);
      A*123+h*123
```

```
>> ausdr;
      A^G*(A+h)
```

In this command the first argument is the expression in which substitution takes place and the substitution itself has to be given in form of an equation. By use of **subs** the expression itself remains unchanged; the substitution is only carried out in a copy of the expression. Substitutions can also be executed subsequently with one command:

```
>> subs(ausdr, A = 123, G = NN);
      123^NN*(h+123)
```

If the roles of **A** and **G** are to be exchanged in **ausdr** the user may be tempted to try:

```
>> subs(ausdr, A=G, G=A);
      A^A*(A+h)
```

However, this does not lead to the desired result since here the substitutions were carried out sequentially: First, **A** was replaced by **G** and then all **G**'s were replaced again by **A**, so that the result does not contain any more **G**'s. In case this sequential execution of substitutions is not desired and a parallel substitution should take place instead, then the substitution equations have to be put in square brackets:

```
>> subs (ausdr, [A=G, G=A]);
      G^A*(G+h)
```

Sometimes it is desirable to replace an identifier in an expression at only one position and to leave it unchanged in all others. In order to do this the function **subsop** is available where the quantities to be substituted are specified by the corresponding paths. For example, if in **ausdr** only the second **A** is to be replaced, then the user should use **subsop**:

```
>> subsop(ausdr, [2,1]=a);
      A^G*(a+h)
```

With this command, several substitutions can be carried out:

```
>> subsop(ausdr, [1,1]=a, [2,1]=aa);
      a^G*(h+aa)
```

An option for parallel execution in **subsop** does not exist. The user should bear in mind that **subsop** always carries out its substitution sequentially:

```
>> A := a^b; subsop(A, [1]=c^d, [1,2]=123);
      a^b
      (c^123)^b
```

although the following example seems to contradict that claim:

```
>> A:=NIL: B:=NIL: ausdr := A^G*(A+h):

>> subsop(ausdr, [2,1]=(B+C));
     A^G*(B+C+h)

>> subsop(ausdr, [2,1]=(B+C), [2,2]=g);
     A^G*(B+C+g)
```

 If the substitutions had been carried out sequentially then the result would have been `A^G*(B+g+h)`. However, the user should observe that the first substitution yields `A^G*((B+C)+h)` and only when the expression is given out does the *simplifier* transform it into normal form `A^G*(B+C+h)`. In order to increase execution speed in the implementation of `subsop` no simplification takes place between substitutions. With this in mind, the user can see in the example above that the second substitution `subsop(ausdr, [2,1]=(B+C), [2,2]=g);` works on `A^G*((B+C)+h)` and yields the correct result `A^G*((B+C)+g)`. Only when returning the expression after the execution of `subsop` does the simplifier leave out the unnecessary brackets.

MuPAD also provides a function for the replacement of expressions which do not form a complete subtree. For this the system function `subsex` is provided:

```
>> subsex(a*b*c, a*b=46);
     c*46
```

As already mentioned `subs` only replaces complete subtrees. Since `a*b` in the example above is not a complete subtree, the desired substitution cannot be carried out by use of `subs`:

```
>> subs(a*b*c, a*b=46);
     a*b*c
```

Unfortunately, `subsex` is considerably slower than `subs` since a more sophisticated search strategy has to be applied.

2.10   Evaluation

In order to understand the evaluation mechanism of MuPAD the user should first look at an assignment, lets say the assignment `A := b*c`. If such an assignment is included in round brackets then it forms a valid MuPAD expression. Normally, expressions are evaluated immediately. In order to prevent immediate evaluation MuPAD provides the environment variable `EVAL_STMT`. If this variable attains the value `FALSE` then the automatic evaluation of assignments in expressions is prevented enabling the user to analyze operands of the assignment `A := b*c` by using the function `op`:

```
>> EVAL_STMT := FALSE:
```

```
>> op((A := b*c), 0..nops((A := b*c)));
   _assign, A, b*c

>> EVAL_STMT := TRUE:
```

Assignments are expressions formed on their highest level by use of the function
`_assign`. The tree structure of such an assignment is as follows:

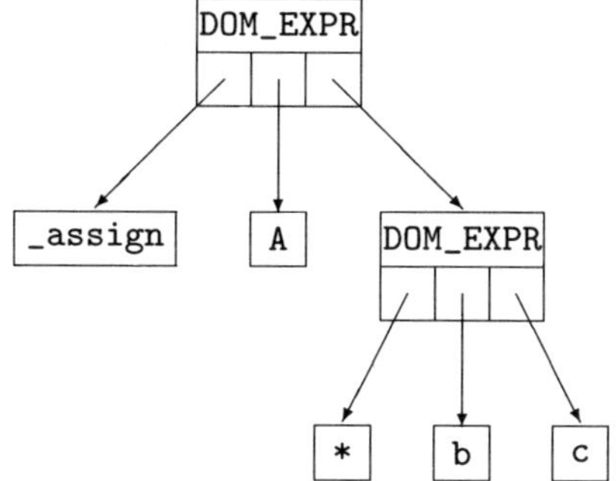

Figure 2.2: Assignment `A := b*c`

If, for a simplified understanding, such an `_assign`-node is symbolized for the
moment by a boldface arrow then the user obtains the diagram depicted in figure
2.3 for the assignment above.

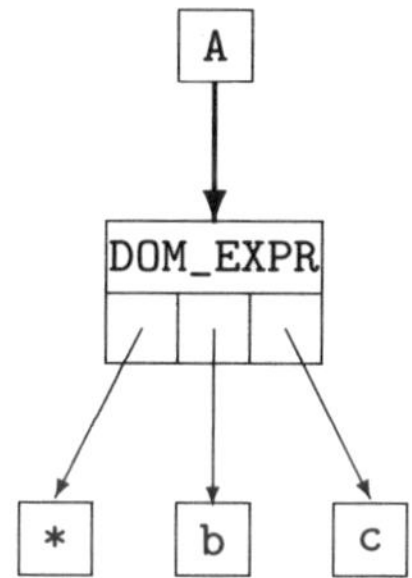

Figure 2.3: Assignment `A := b*c`

The repeated assignment `A := b*c; b := e; e := f;` then leads to figure 2.4.
On calling `A`, MuPAD follows the boldface arrows. The total number of boldface
arrows MuPAD runs through while evaluating `A` is called the *substitution depth*.
In this example the substitution depth is 3. While working interactively with the
system, all substitutions according to the boldface arrows are normally carried out.
This, however, only to a maximal depth as defined by the environment variable

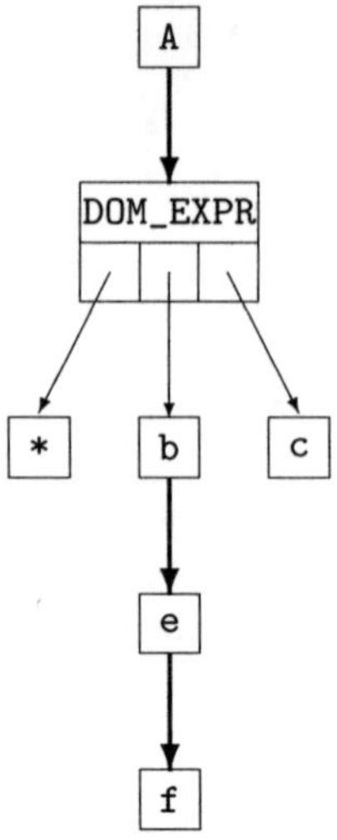

Figure 2.4: Assignments `A:=b*c; b:=e; e:=f;`

LEVEL. The current value of level is easily found:

```
>> LEVEL;
       100
```

In order to prevent a recursive evaluation resulting in an infinite loop there is a second environment variable **MAXLEVEL**:

```
>> MAXLEVEL;
       100
```

which, in general, has the same value as **LEVEL**. If, during evaluation, this value is reached, then the system assumes that there is a recursive assignment:

```
>> B := C: C := B:
```

```
>> B;
       Error: Recursive Definition
```

However, this need not necessarily be true as the following example performing 100 assignments `b1 := b2:  b2 := b3:  ...  b99 := b100:` shows:

```
>> (b.i := b.(i+1)) $ i=1..100:
```

```
>> b1;
       Error: Recursive Definition
```

Calling `b1` does indeed lead to a substitution depth of 100. By this **MAXLEVEL** is reached and MuPAD erroneously assumes a recursive definition. If the value of

LEVEL and MAXLEVEL had been 101 then calling b1 would have led to the correct output of b100.

In case the value of LEVEL is less than that of MAXLEVEL then recursive definitions are never detected:

```
>> LEVEL := 99: MAXLEVEL := 100: B := C: C := B:

>> B;
     C
```

However, the user is urgently warned against tampering with the environment variables LEVEL and MAXLEVEL. We have only dealt with them here in order to explain how evaluation works in MuPAD. 

MuPAD provides the function level which, with respect to evaluations, has the same effect as a temporary change of the environment variable LEVEL. The calling of level(expr, n) assigns locally, i.e., only during the evaluation of the expression expr, the value n to the environment variable LEVEL. This is seen in the following example where the environment variables LEVEL and MAXLEVEL have been set back to their original default values:

```
>> a := b: b := a:

>> level(a, 3);
     b

>> level(a, 4);
     b

>> level(a, MAXLEVEL);
     Error: Recursive Definition
```

The skillful use of the options available for evaluation provides an efficient tool for programming. Often the execution speed of programs written in MuPAD increases considerably if unnecessary evaluations are prevented. It should be mentioned that the call level(expr) results in a complete evaluation, thus this call is equivalent to level(expr, MAXLEVEL).

In addition there is a function val, which evaluates to the substitution depth 1 but, contrary to level, it does not simplify its result. The difference between val(expr) and level(expr, 1) becomes clear in the following example:

```
>> reset();

>> a := b: b := 1:

>> val(a+b+2);
     b+1+2
```

```
>> level(a+b+2, 1);
     b+3
```

To suppress undesired evaluations there is another instrument, the function `hold`, which prevents all evaluations:

```
>> sqrt(4);
     2
```

```
>> hold(sqrt(4));
     sqrt(4)
```

```
>> hold(1+2);
     1+2
```

At first it is surprising that in this example a subsequent `level` does not change the result:

```
>> hold(1+2);
     1+2
```

```
>> level(%);
     1+2
```

This somewhat confusing behavior is caused by the fact that when calling up the history table's entries no evaluation takes place. In order to obtain the desired result 3, the user needs a function which causes a subsequent evaluation of those system functions which, like `hold` or `last`, return their results in unevaluated form. This functionality is provided by the system function `eval`:

```
>> reset():
```

```
>> a := b: b := c: c := d;
     d
```

```
>> hold(a);
     a
```

```
>> eval(%);
     d
```

2.11   Programming

The most important aim in the development of programming languages is to provide the user with the possibility of writing his own procedures. Computer algebra systems have the advantage that procedures can be written in a very short way.

This is due to the efficient options for data manipulation. In MuPAD this is especially true, since a wide variety of tools for data manipulation are provided. Generally, procedures are expressions assigned to identifiers. If such procedure assigned identifiers are called with suitable arguments then, if programmed correctly, the evaluation of the procedure will lead to the desired result. First we want to analyze a simple procedure which returns the product of its arguments. We assign this procedure to the identifier f, i.e. we want to have a program that does the following $f(x, y) = x * y$. However, this definition cannot be used as a valid program since the system would consider this expression an equation:

```
>> reset();
```

```
>> f(x,y)=x*y;
      f(x, y)=x*y
```

```
>> type(%);
       "_equal"
```

However, in the file "tutorial" we have provided a procedure which allows a similar functional definition for programs:

```
>> FUNC(f(x,y)=x*y):
```

```
>> f(2, 3); f(91233, 778);
      6
      70979274
```

If the identifier f is called now, then the user obtains a first look at the structure of a MuPAD procedure:

```
>> f;
    proc(x, y)
    name f;
    begin
      x*y
    end_proc
```

The procedure f starts with the keyword **proc** which is followed immediately by the formal parameters of the procedure given in brackets, then by the name of the procedure. Next, starting with the keyword **begin**, the algorithm starts. The end of the procedure is marked again by a keyword, namely, **end_proc**. The algorithm between the keywords **begin** and **end_proc** is called the procedure body. In this case the procedure body consists of a single statement. Indeed, the same procedure could have been defined by interactive input of the expression sequence consisting of the corresponding keywords and statements:

```
>> h := proc(x, y)
      begin
        x*y
      end_proc:
```

```
>> h(456, 776);
     353856
```

Here the user should keep in mind that the assignment of the procedure to the identifier **h** has to be marked by := and that this assignment, as all MuPAD commands, must be ended by either a semicolon or a colon.

If the operands of **f** are now investigated:

```
>> op(f);
     (x, y), NIL, NIL, x*y, NIL, f
```

It is seen that a procedure generally consists of six operands. The fourth operand is the algorithm, i.e. the most important part given by the statement sequence which forms the procedure body. Of the six possible operands in our example only three were used, therefore the others are represented by the entry **NIL**. Obviously the first operand is the expression sequence defined by the formal parameters. As already mentioned, the fourth operand is given by the statements of the procedure body, and in the sixth operand the name of the procedure is stored. Further statements do not increase the number of operands. This is seen by looking at the following example, where the square of the product of arguments is calculated:

```
>> quad := proc(x, y)
          begin
            x*y;
            %^2
          end_proc:

>> quad(2, 3);
     36

>> op(quad);
     (x, y), NIL, NIL, (x*y; last(1)^2), NIL,   quad
```

Again, all statements forming the procedure body are contained in one operand. Of course, the statements in the procedure body can be accessed by use of the function **op** and the corresponding path names:

```
>> op(quad, [4, 1]);
     x*y

>> op(quad, [4, 2]);
     last(1)^2
```

All identifiers can be used as formal parameters for procedures. For this it is insignificant whether a value has been assigned to their names outside the procedure:

```
>> x := 15: FUNC(g(x)=x^10):
```

```
>> g(2);
      1024
```

For computations in programs identifiers can be used in order to store intermediate results. So, instead of `quad` the following modification can be used:

```
>> quad_mod := proc(x, y)
              begin
                A := x*y;
                A^2
              end_proc:
>> quad_mod(2, 3);
      36

>> A;
      6
```

However, this program has a side effect, namely, that after its execution a value is assigned to the identifier **A**. Hence the identifier **A** was considered a *global variable* and now has a new value. If, in some other context, **A** had been used to keep track of another value, this information is lost. Therefore, wherever possible, global variables should not be used. There is the possibility that assignments to identifiers are only valid within the context of the program. Then their meaning outside that procedure is not effected. These variables are called *local variables* and they have to be declared in a special way:

```
>> quad_better := proc(x, y)
                 local b, c;
                 begin
                   b := x*y;
                   b^2
                 end_proc:
>> quad_better(2, 3);
      36

>> b; c;
      b
      c
```

As seen here, several local variables can be declared, the user only has to write them one after another: `local b, c;`. It is not important whether or not these variables are really needed within the program. The expression sequence, consisting of the local variables, forms the second operand of the procedure:

```
>> op(quad_better, 2);
      b, c
```

Thus, local variables are those variables which can be used within a procedure without changing their meaning outside that procedure. The values which they

obtain during the execution of the procedure are forgotten after the procedure has
been executed. So, the value 15 can be assigned to the identifier a and the user
can still work with other assignments a := 2323243545 within the procedure, the
user only has to declare a as a local variable:

```
>> a := 15:
```

```
>> test := proc(e)
          local a;
          begin
            a := 2323243545; a*e
          end_proc:
```

```
>> test(s); a;
     s*2323243545
     15
```

After execution of the procedure, a has the value 15 again.

We already know a number of environment variables like DIGITS, TEXTWIDTH,
LEVEL or PRETTY_PRINT. An interesting feature of MuPAD is that these environment variables can also be declared as local variables. As is the case with all other
local variables, assignments within procedures are only effective in the context of
that procedure. The only difference is that these environment variables are automatically initialized even if they are used as local variables. Their initial value is
the one they have in the environment from which the procedure was called:

```
>> illustrate_en_var :=  proc(x)
                    local DIGITS;
                    begin
                      print(float(x));
                      DIGITS := 50;
                      print(float(x))
                    end_proc:
```

```
>> illustrate_en_var(1/3);
     0.3333333333
     0.33333333333333333333333333333333333333333333333333
```

```
>> DIGITS;
     10
```

```
>> float(1/3);
     0.3333333333
```

In this example, the user should observe that calling float(x) instead of the
second call of print(float(x)) within the procedure illustrate_en_var only

leads to a ten-digit return of the decimal number for 1/3. The reason for this is that output only takes place after a procedure is quit, i.e., at a time when the variable DIGITS has its default value again.

The third operand in a procedure is formed by the expression sequence comprising of the so-called *options*. Options are parameters for controlling procedures. At present only the options hold and remember are available. The option hold prevents evaluation of the arguments of a procedure. This can be seen in the following example:

```
>> a := 0:

>> illustrate_hold := proc (b) option hold;
                    begin
                      b
                    end_proc:

>> illustrate_hold(a);
     a

>> a := 0: no_hold := proc (b) begin  b end_proc:

>> no_hold(a);
     0
```

In the first case only evaluation of the call illustrate_hold(a) with a subsequent eval leads to the expected result:

```
>> illustrate_hold(a); eval(%);
     a
     0
```

If the option remember is given then MuPAD stores all results computed with that procedure. These stored results can then be found in a table, the so called remember table, which forms the fifth operand of a procedure.

MuPAD accesses values in the remember table directly. The values computed by the procedure can be found in the remember table:

```
>> sin_quad := proc(x)
              option remember;
            begin
              float(sin(x)^2)
            end_proc:

>> sin_quad(3); sin_quad(PI); sin_quad(1.1);
     0.1991485667e-1
     0.0
     0.7942505586
```

```
>> op(sin_quad, 5);
      table(
           PI = 0.0,
           3 = 0.1991485667e-1,
           1.100000000 = 0.7942505586
           )
```

 Procedures with side effects, for example those using global variables, can lead to erroneous results if the option **remember** is used:

```
>> Proc_with_side_effect := proc(i)
                                 option remember;
                             begin
                                 A*i
                             end_proc:
```

```
>> A := 15: Proc_with_side_effect(5);
      75
```

```
>> A := 25: Proc_with_side_effect(5);
      75
```

Values can also be inserted in remember tables without using the option **remember**. Of course, this mechanism can also be used to insert absolute nonsense in the remember tables, so that the entries have nothing to do with the algorithm used in the procedure body. If nonsense is entered in the remember table this is given out in future calls of the procedure:

```
>> sin_quad(PI) := "wrong result": sin_quad(PI);
      "wrong result"
```

If an argument has a value already assigned in the remember table, then any new entry for that argument replaces the old one.

It should be mentioned that a functional assignment to an unassigned identifier leads to the creation of a procedure:

```
>> reset():
```

```
>> ff(1) := 23234;
      23234
```

```
>> ff;
      proc ( )
      name ff;
      option remember;
      begin
         procname( args( ) )
      end_proc
```

```
>> op(ff, 5);
    table(
        1 = 23234
        )
```

This procedure then, apart from the single entry in the remember table, only returns its evaluated argument:

```
>> ff(z);
    ff(z)
```

```
>> ff(1);
    23234
```

In this case the option **remember** is set automatically.

A further possibility of defining procedures is given by the underline function `_procdef`:

```
>> f := _procdef((x, y), (a, b), hold, x^y, NIL, f);
    proc(x, y)
    name f;
    local a, b;
    option hold;
    begin
      x^y
    end_proc
```

However, the inexperienced user is warned against using this possibility. Generally he should avoid using any underline function. A wrong use of these can, as we will explain in section 2.16, lead to a breakdown of the system. 

Procedures can call each other. Consider, for example, the following two procedures:

```
>> f := proc(x)
       begin
         sin(x)
       end_proc:
```

```
>> fdiff := proc(x) local y;
           begin
             subs(diff(f(y), y), [y = x])
           end_proc:
```

The first computes the **sine** of its argument, whereas the second computes the derivative of **f** at **y** and then replaces **y** with the argument passed to that procedure. Here **diff** is the built-in differential function of MuPAD. As seen here the second procedure calls the first one:

```
>> fdiff(x);
     cos(x)
```

```
>> fdiff(1.5);
     0.7073720166e-1
```

Within procedures the user can work with MuPAD data in the usual way. However, there is one essential difference: generally, in procedures only a one-level substitution takes place! By this behavior execution speed of procedures is increased considerably. If other evaluation depths are desired, then this can be achieved by an explicit use of the system function **level** or by redefining **LEVEL** within the procedure:

```
>> eval_1level := proc()
              begin
                 a := b;
                 b := c;
                 c := d;
                 a
              end_proc:
```

```
>> eval_1level();
     b
```

If the same assignments are given interactively, then a complete evaluation takes place:

```
>> a := b:    b := c:    c := d:    a;
     d
```

If, in the last procedure, a complete evaluation of the final **a** is desired then **level** should be used:

```
>> eval_all_level := proc()
                 begin
                    a := b;
                    b := c;
                    c := d;
                    level(a)
                 end_proc:
```

```
>> eval_all_level();
     d
```

To facilitate programming MuPAD offers a wide variety of control structures. Among them are:

The `if`-statement:

```
if < condition > then
        < statement >
{ else
         < statement > }
end_if

Example:
if a > 4 then a else 4 end_if
```

The `for`-loop :

```
for < running index > from < begin > to < end >
      {step < step-width >} do
         < statement >
end_for

Example:
for i from 12 to 32 step 2 do a := (a, i) end_for
```

The `while`-loop :

```
while < condition > do
        < statement >
end_while

Example:
while a < 6 do a := a + 1 end_while
```

The `repeat`-statement :

```
repeat
        < statement >
until < condition >
end_repeat

Example:
repeat a := a + 1 until a > 6 end_repeat
```

In these tables all places where an entry has to be given are designated by pointed brackets $<\ >$ whereas the optional parts are marked by braces $\{\ \}$. If the commands in these examples are executed then this leads to:

```
>> a := 2: if a>4 then a else 4 end_if;
   4

>> a := 34: if a>4 then a else 4 end_if;
   34

>> a := (beginning):

>> for i from 12 to 32 step 2 do a := (a, i) end_for;
   beginning, 12, 14, 16, 18, 20, 22, 24, 26, 28, 30, 32

>> a := -3: while a<6 do a := a + 1 end_while;
   6

>> a := -3: repeat a := a + 1 until a>6 end_repeat;
   7
```

If the **repeat**-statement is used the execution runs through the body of the loop at least once. This is because **repeat** executes its statement before the given condition is checked.

The list of control structures given here is by no means complete, there are a number of modifications. In the **if**-statement we should mention the **elif** and in the **for**-loop the user can use the **downto** instead of the **to** in order to count in the opposite direction. Furthermore, instead of using an index, a **for**-statement can be made to run through the range given by the operands of an expression.

The **if-elif**-statement :

```
if < condition > then
        < statement >
{elif < condition > then < statement > }
{further elif's}
{else
        < statement >}
end_if

Example:
if a > 4 then a elif a < 3 then "hallo" else 4 end_if
```

The **for**-**in**-loop :

```
for < running index > in < expression > do
        < statement >
end_for

Example:
for i in A*B*C do print(i^2) end_for
```

The **for**-**downto**-loop :

```
for < running index > from < Beginning >
     downto < end > {step < stepwidth >} do
        < statement >
end_for

Example:
for i from 32 downto 12 step 2 do a := (a, i) end_for
```

Executing these examples results in:

```
>> a := 11:

>> if a < 4 then a elif a < 3 then "hallo" else 4 end_if;
     "hallo"

>> for i in A*B*C do print(i^2) end_for;
     A^2
     B^2
     C^2

>> i := NIL: a := (beginning):

>> for i from 32 downto 12 step 2 do a := (a, i) end_for;
     beginning, 32, 30, 28, 26, 24, 22, 20, 18, 16, 14, 12
```

Within loops the special keywords **break** and **next** can be used in order to quit a loop or to jump to the next execution within the same loop:

```
>> for i from 1 to 30 do i; break end_for;
     1

>> a := 2:
```

```
>> while a < 30 do a := a+1; break end_while;
     3

>> a := 2:

>> while a<7 do
     a := a+1;
     if a<5 then
       next
     end_if;
     print(a)
   end_while;
     5
     6
     7
```

The use of **break** leads to unconditionally quitting the loop and by use of **next**
all further statements within the actual loop are ignored and a jump to the next
execution takes place. Note that this implies updating the loop index and checking
for termination.

Within procedures the meaning of these keywords is similar:

```
>> illustrate_break := proc(x)
                      begin
                        print(x);
                        break;
                        "hallo"
                      end_proc:

>> illustrate_break(x);
     x
```

A further useful programming construct is the **case**-statement. This is particularly
efficient because, contrary to a repeated **if**-statement, the expression used for the
comparison is evaluated only once.

The `case`-statement :

```
case <expression>
        of <comparison quantity1> do <statement 1>;
        {of < comparison quantity2> do <statement 2>;}
        {further of's }
        {otherwise <alternative statement >}
end_case

Example:
case a
        of 1 do print(1);
        of 2 do print(2);
        of 3 do print(3)
        otherwise print("no case")
end_case
```

```
>> illustrate_case := proc(a)
                  begin
                    case a
                      of 1 do print(1)
                      of 2 do print(2)
                      of 3 do print(3)
                      otherwise print("no case")
                    end_case
                  end_proc:
```

```
>> illustrate_case(5);
     "no case"
```

However, the following result might be surprising:

```
>> illustrate_case(1);
     1
     2
     3
     "no case"
```

But, this is confusing at first sight only. The effect of the `case`-statement is such, that, after a first affirmative outcome of the comparison, the remaining statements are carried out without any further comparisons. If this is not desired, then, each statement must be followd by a `break`:

```
>> illustrate_case := proc(a)
                  begin
                    case a
                      of 1 do print(1); break
                      of 2 do print(2); break
                      of 3 do print(3); break
                      otherwise print("no case")
                    end_case
                  end_proc:

>> illustrate_case(1);
     1
```

The keyword **next** can be used in exactly the same way it is used in a loop. In order to understand the effect of the keywords **next** and **break** the user should look at the following example:

```
>> probe :=  proc (a)
          begin
            case a
              of 1 do print( 1 ); break
              of 2 do print( 2 ); next
              of 3 do print( 3 )
              of 2 do print( 2 ); break
              of 4 do print( 4 )
            end_case
          end_proc:
```

With different arguments the user obtains:

```
>> probe(1);
     1

>> probe(2);
     2
     2

>> probe(3);
     3
     2

>> probe(4);
     4

>> probe(5);
```

Here, on evaluation of the **break**, the **case**-statement is quit. On evaluation of **next** the remaining **case**-statement is processed as if it were a new **case**-statement. If neither of the keywords **break** or **next** is used then the remaining statements are executed without performing any comparisons.

Any **case**-statement can be replaced by a corresponding, logically equivalent **if**-statement. For example:

```
>> case_proc := proc (a)
              begin
                case a
                  of 0 do 0; break
                  of 1 do 1; break
                  of 2 do 2; break
                  of 3 do 3; break
                end_case
              end_proc:
```

is obviously equivalent to:

```
>> if_proc := proc ( a )
            begin
              if a = 0 then
                0
              elif a = 1 then
                1
              elif a = 2 then
                2
              elif a = 3 then
                3
              end_if
            end_proc:
```

Nevertheless, the **case**-statement should be preferred, since it is executed much faster.

2.12 Input, Output, Files and Text

For the control of output on the screen we are already acquainted with the environment variables **TEXTWIDTH**, **PRETTY_PRINT**, and the function **print**. Apart from these, there are further functions providing more functionality with respect to in- and output of files and for writing protocols of sessions. In this respect, MuPAD offers a convenient variety of possibilities to the user.

If, during the execution of a program, a value is to be read, then this can be done by use of the functions **input** or **textinput**:

```
>> input("Please, insert value for A", A);
      Please, insert value for A>> 1234:

>> A;
   1234
```

The use of `input("message",A)` first prints the `message` on screen as output followed by a prompt. Until the input, which is required at this point, is terminated the program stops. The interactive input is then assigned, as a value, to the identifier `A`. Input has to be ended by a colon or a semicolon and is terminated by either <Return> or <Enter>, depending on the hardware platform. After termination the program continues to work with this value assigned to the identifier `A`. The function `textinput` has a similar function. Here, however, input is automatically considered a string:

```
>> textinput("Please insert string as value for B", B):
      Please insert string as value for B>>example string

>> B;
    "example string"
```

Again a prerecorded message is given as output and the input is assigned, as a value to the identifier `B`. Since the input is automatically considered as a string, no quotes need to be given. Quotes within a text are considered as characters, therefore, they do not have to be masked by a backslash. Apart from the different interpretation of the given input as a string, there is another slight difference in the use of `input` and `textinput`. For the application of newlines in `textinput`, <Return> and <Enter> are interpreted in the usual way, so they cannot be used for termination of textual input. For this purpose <Ctrl-D> has to be given.

If text and data have to be read from files, the user can use the functions `read`, `finput` and `ftextinput`. To store text, MuPAD data or the special M-code data in files the functions `protocol`, `write` and `fprint` are provided:

```
>> A := 124: B := "hallo":

>> write(Text, "filename", A, B);

>> A := 0;
   0

>> read("filename"): A;
   124
```

The function `write` writes into the file with the name given as a string in the call of the `write` command; in this case the file with the name `"filename"`. The option `Text` ensures that the data are written in text format, without this option another format, the so-called M-code is used. The command `read` reads MuPAD data from a file and executes it. So, in the example above, the variable A, although a new value has been assigned to it in the meantime, has its original value again after execution of the **read** command. The function **ftextinput** reads text line by line. With these utilities the user can now easily write a procedure for reading in a complete text, for example:

```
>> readtext := proc(fname)
              local fid, text, line;
              begin
                if args(0) <> 1 then
                  error("wrong no of args")
                end_if;
                if domtype(fname) <> DOM_STRING then
                  error("no filename")
                end_if;
                fid:= fopen(fname);
                text:= ftextinput(fid);
                if text = null() then return("")
                end_if;
                while (line:= ftextinput(fid)) <> null() do
                  text:= text . "\n" . line
                end_while;
                text
              end_proc:
```

```
>> readtext("filename");
     "A:=hold(124):\nB:=hold(\"hallo\"):"
```

This function, **readtext**, considers the file contents a string and returns it in quotes. This also allows the user to see which form the data were saved in by **write**. Interesting, at this point, is that all assignments, when written as text, were saved with a **hold**. This **hold** ensures that the complete assignment structure of a session is saved. The user should observe that **write** overwrites the contents of the corresponding file. If data are to be appended to a file, then the user has to proceed as described below. Instead of **write**, **fprint** can also be used. Then the data are written in the file the same way as if they were printed on screen by use of the **print** command. The command **write** is used to save the state of variables during a session, including the assignment structure, whereas **fprint** is used for saving their value.

If **Bin** is given as the first parameter in **write**, or if there is only one parameter, then the MuPAD data are saved as M-code. This code, although independent of the hardware platform, preserves the internal tree structure of the data. When reading such code with **read** or **finput** MuPAD automatically recognizes that these data already reflect the tree structure and, therefore, need not be processed by the parser. In this case the reading of data is performed very quickly since parsing and evaluation are unnecessary.

To protocol a session, the user has to save output and input on screen. To do this, the system function `protocol` is used. By `protocol("filename")` the session is saved in a file with the name `"filename"`.

The protocol is ended by giving either `protocol()` or `protocol("newfile")`. The last command, however, results in a new protocol being written in `"newfile"`:

```
>> protocol("newfile"):

>> "now do something with MuPAD";
     "now do something with MuPAD"

>> 2*8*9927726265;
     158843620240

>> protocol():

>> readtext("newfile");
 " \"now do something with MuPAD\";\n\"now do something with MuPAD\
 \"\n\n 2*8*9927726265;\n158843620240\n\n protocol():"

>> print(Unquoted,%);
        "now do something with MuPAD";
        "now do something with MuPAD"

          2*8*9927726265;
          158843620240

          protocol():
```

Here we see that in general text is printed in the format of the data type **string** with quotes marked by backslashes and linefeeds given as \n. However, in the commands **print** and **fprint** this format can be avoided by using the option **Unquoted**.

If the user wants to read data from a file,

$$finput("filename", identifier)$$

or

$$ftextinput("filename", identifier)$$

should be used instead of the functions **input** and **textinput** which were described above. The second parameter given here is the identifier to which the data are going to be assigned. When these two functions are used no message is given on screen. Again the function **finput** automatically recognizes whether the data were saved in M-code or not.

Apart from this command, MuPAD possesses further possibilities of working with files. These are given by the functions **fopen** and **fclose**. **fopen** accepts up to three parameters. These can occur in the following combinations:

```
fopen("newfile"):
fopen("newfile", Write):
fopen("newfile", Append):
fopen(Bin, "newfile", Write):
fopen(Bin, "newfile", Append):
fopen(Text, "newfile", Write):
fopen(Text, "newfile", Append):
```

If executed successfully, i.e. if opening the file **"newfile"** was successful, the return of this command is an integer > 0. This number, the so-called *file descriptor*, can be used in subsequent commands. The mode **Write** causes the corresponding file to be overwritten, whereas **Append** adds to the current contents of that file. **Bin** causes data to be saved in M-code. If instead of this **Text** is used then the data are saved in text format. If only one parameter is given in **fopen** then the corresponding file can only be read. In this case MuPAD recognizes if the data are text or M-code.

For processing text there are several functions:

```
>> A := "This is a short text";
   "This is a short text"

>> write(Text, "newfile", A);

>> B := readtext("newfile"):

>> B; print(Unquoted, B);
   "A:=hold(\"This is a short text\"):"
   A:=hold("This is a short text"):

>> text2tbl(B, [" "]);
   table(
        1="A:=hold(\"This",
        2=" ",
        3="is",
        4=" ",
        5="a",
        6=" ",
```

```
        7="short",
        8=" ",
        9="text\"):"
)
```

Here, the assignment to A was written as a MuPAD datum into the file `"newfile"`.
Then, it was read in as a string by `readtext` and assigned to the identifier B. By
use of the conversion command `text2tbl` (text-to-table) the text was broken up
at the separation points defined in this command, here, at the blanks (`" "`). The
resulting substrings were then written as entries in a table. From here further
processing of text pieces can take place. The system function `text2list` works in
a similar way, but here the text sections are saved in a list instead of a table. The
function `text2list` is usually faster than `text2tbl`. In these functions, several
separation marks can be given:

```
>> text2tbl(B, [" ", "s"]);
    table(
         1 = "A:=hold(\"Thi",
         2 = "s",
         3 = "",
         4 = " ",
         5 = "i",
         6 = "s",
         7 = "",
         8 = " ",
         9 = "a",
        10 = " ",
        11 = "",
        12 = "s",
        13 = "hort",
        14 = " ",
        15 = "text\"):"
    )
```

Here the user has the additional possibility of running through the separation
marks in a cycle. This means that separation marks, after having been used, can
only be used again after all the remaining marks have been used too:

```
>> text2tbl(B, [" ", "r"], Cyclic);
    table(
         1 = "A:=hold(\"This",
         2 = " ",
         3 = "is a sho",
         4 = "r",
         5 = "t",
         6 = " ",
```

```
        7 = "text\"):"
   )
```

Text, which without quotes would be a meaningful MuPAD datum, is easily converted to such:

```
>> text2expr("AA := x+y");
    AA := x+y
```

`text2expr` does not evaluate the datum as can be seen by:

```
>> AA;
    AA
```

For this purpose the user has to use an explicit call of `eval`:

```
>> eval(text2expr("AA := x+y"));
    x+y

>> AA;
    x+y
```

2.13 Graphics

MuPAD has very comfortable graphic facilities. Plots necessary for graphical representation of mathematical objects can be obtained by working interactively with the graphics tool VCam. There, as briefly described in section 5.1.4 of this tutorial, self explaining buttons and menus will easily and comfortably navigate even the less experienced user. But every plot which can be obtained interactively, can also be achieved by calling the corresponding MuPAD commands. These commands, as well as the different plot styles and options made available by the MuPAD language, are illustrated in the following. We will not give a systematic introduction into this subject but rather prefer to present a variety of examples instead.

To plot a curve we use the `plot2d` command:

```
>> plot2d(Axes = Origin, [Mode = Curve,
    [u, sin(u)], u = [-PI, PI],
    Grid = [40]
    ]);
```

where we have to specify the parametrization, the range of the chosen parameter u and the number of points to be calculated (here 40). In future MuPAD releases there will be other two dimensional graphic objects, therefore, the graphic command also needs the specification that the plot shall be a `Curve` under `Mode`. The result of this plot command is:

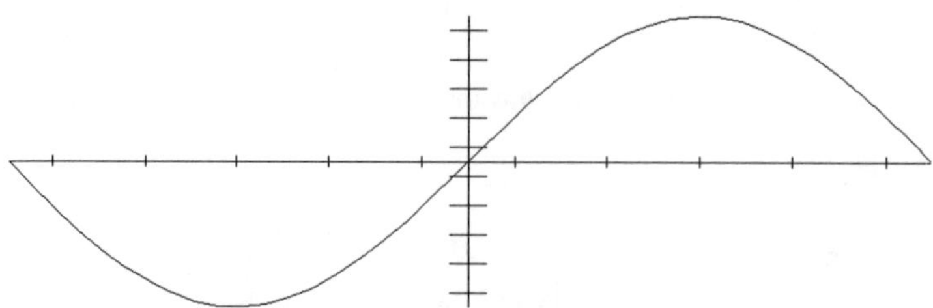

As normally done for univariate functions, the x-variable coincides with the curve
parameter. However, other parametrizations can also be chosen as seen by the
following plot of a spiral:

```
>> plot2d(Axes = Corner, [Mode = Curve,
   [u*cos(u), u*sin(u)], u = [0, 2*PI],
   Grid = [40]
   ]);
```

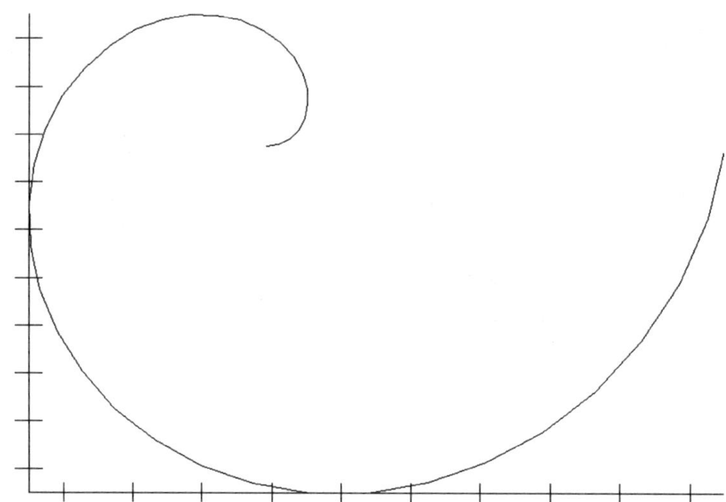

If axes are not desired, then, optionally, we can make the specification **Axes =
None**:

```
>> plot2d(Axes = None,
   [Mode = Curve,
   [sin(u)*sin(2*u)*sin(3*u),
    0.5*sin(4*u)*sin(5*u)*sin(6*u)],
```

```
u = [0, 2*PI], Grid = [50], Smoothness = [5],
Style = [LinesPoints]
]);
```

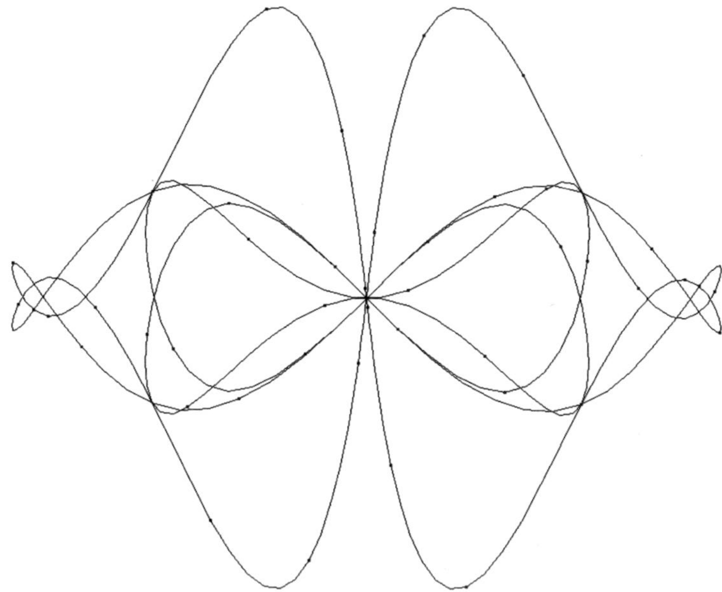

In this example we have also become acquainted with a new style for the representation of curves, the [LinesPoints] style. Here the sample points are marked by dots and are connected by straight lines. Depending on the curvature of the curve the result of the linear interpolation might be not smooth enough. In this case the parameter Smoothness can be set to an integer n $\geq$ 1. Then n intermediate points between two sample points are calculated and used for the interpolation of these two sample points. As a result the curve will now be much smoother.

We can define the coordinates of Pascal's snail and parameter-dependent epicycles with:

```
>> pascal_x := proc(a, l, u)
   begin
     a*cos(u)^2 + l*cos(u):
   end_proc:

pascal_y := proc(a, l, u)
   begin
     a*cos(u)*sin(u) + l*cos(u):
```

```
      end_proc:
epiz_x := proc(a, b, u)
    begin
      (a+b)*cos(u) - a*cos((a+b)*u/a):
    end_proc:

epiz_y := proc(a, b, u)
    begin
      (a+b)*sin(u) + a*sin((a+b)*u/a):
    end_proc:
```

MuPAD allows to plot several curves into the same window, for example as in:

```
>> plot2d(Axes = Corner, Ticks = 0,
    [Mode = Curve,
        [hold(pascal_x(1, 1, u)),
         hold(pascal_y(1, 1, u))],
        u = [0, 2*PI], Grid = [50],
        Smoothness = [1],
        Style = [Points]],
    [Mode = Curve,
        [hold(pascal_x(1.5, 1, u)),
         hold(pascal_y(1.5, 1, u))],
        u = [0, 2*PI], Grid = [50],
        Smoothness = [1],
        Style = [Lines]],
    [Mode = Curve,
        [hold(pascal_x(2, 1, u)),
         hold(pascal_y(2, 1, u))],
        u = [0, 2*PI], Grid = [50],
        Smoothness = [1],
        Style = [LinesPoints]
    ]);
```

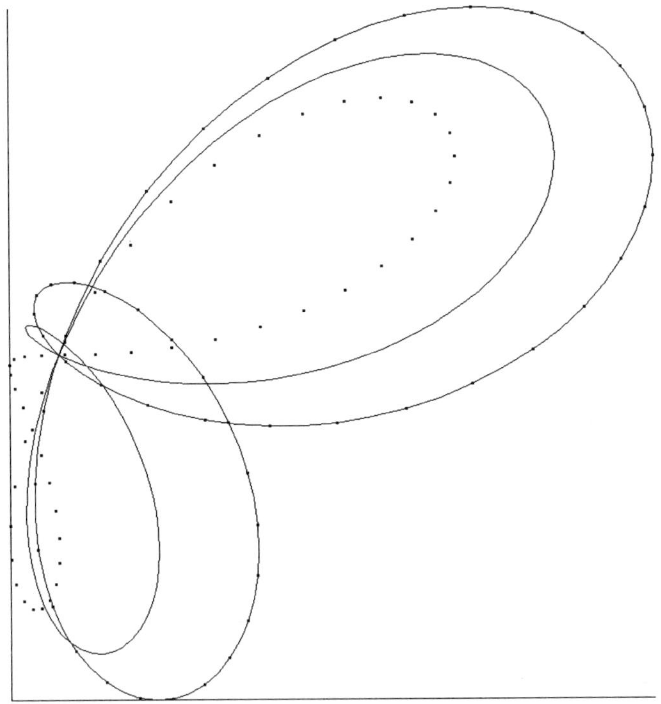

or as in:

```
>> plot2d([Mode = Curve,
        [hold(epiz_x(0.2, 3, u)),
         hold(epiz_y(0.2, 3, u))],
        u = [0, 2*PI], Grid = [50], Smoothness = [5]],
   [Mode = Curve,
        [hold(epiz_x(0.6, 3, u)),
         hold(epiz_y(0.6, 3, u))],
        u = [0, 2*PI], Grid = [50], Smoothness = [5]],
   [Mode = Curve,
        [hold(epiz_x(1.0, 3, u)),
         hold(epiz_y(1.0, 3, u))],
        u = [0, 2*PI], Grid = [50], Smoothness = [5]
   ]);
```

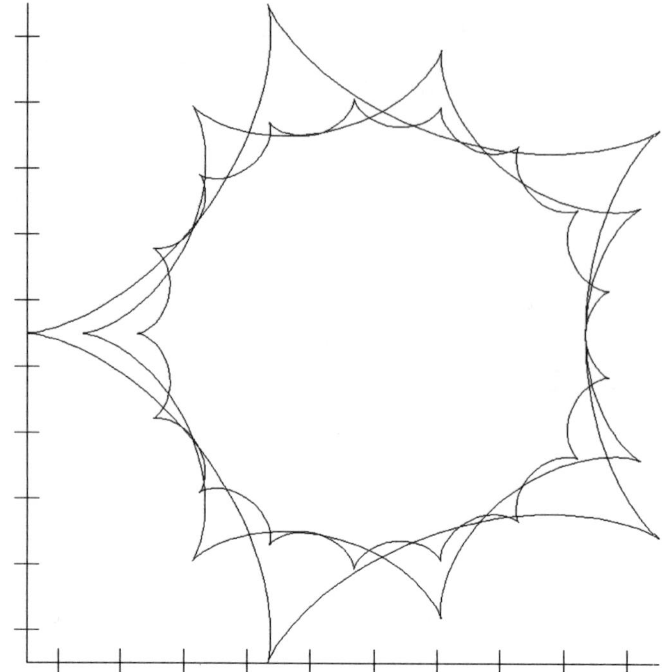

For the second example we had to increase the **Smoothness** considerably in order
to give a correct representation of the points of return of the epicycles. Of course,
for different graphic objects different styles may be chosen. This can be seen in the
first of the two examples above, where three of the four available styles **Points**,
Lines, **LinesPoints** and **Impulses** were used.

As seen in the following plot of a damped oscillation with envelopes, explicitly
specifying the ticks by **Ticks = 0** suppresses the marks at the axes by which the
plot is usually framed:

```
>> plot2d(Axes = Corner, Ticks = 0,
   [Mode = Curve,
        [u, 3*exp(-u/3)*sin(3*u)],
        u = [0, 2*PI], Grid = [50],
        Smoothness = [1]],
   [Mode = Curve,
        [u, 3*exp(-u/3)],
        u = [0, 2*PI], Grid = [50]],
   [Mode = Curve,
        [u, -3*exp(-u/3)],
        u = [0, 2*PI], Grid = [50]
   ]);
```

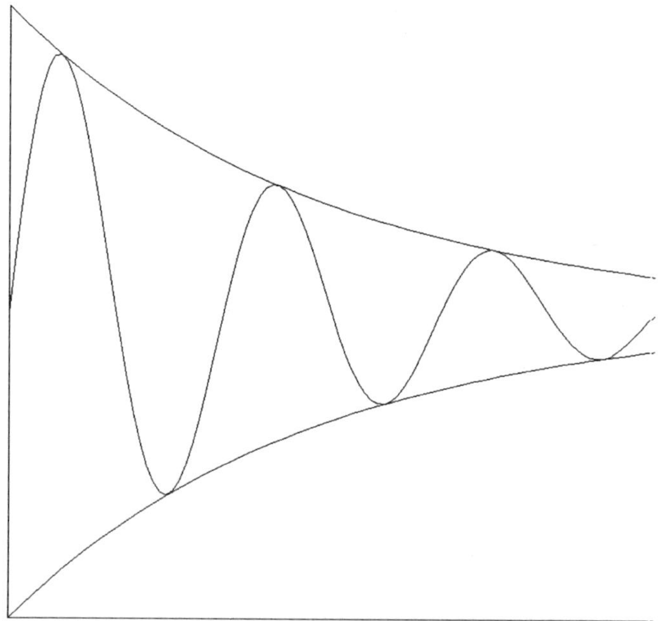

MuPAD can also plot three-dimensional graphics. This is seen in the following examples whereby the first two represent soliton solutions of the nonlinear Korteweg-de Vries equation, and where in the last a third dimension is added to a family of epicycles:

```
>> kdv := proc(u, v, c)
   begin
       e     := exp(c*u -c^3*v):
       f     := 1 + e;
       fx    := c*e;
       fxx   := c*fx;
       3*(fxx/f - (fx/f)^2);
   end_proc:

   plot3d(Axes = Box, Ticks = 0,
          [Mode = Surface,
               [u, v, hold(20*kdv(u, v, 3/5))],
               u = [-10, 10], v = [-10, 10],
               Grid = [40, 30], Style = [HiddenLine, VLine]
          ]);
```

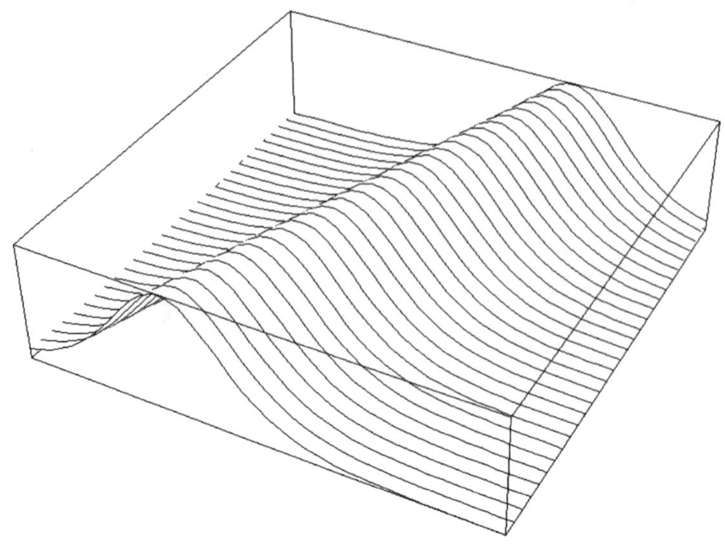

```
>> kdv := proc(u, v, c1, c2)
      begin
        e1     := exp(c1*u - c1^3*v);
        e2     := exp(c2*u - c2^3*v);
        c12    := ((c1 - c2)/(c1 + c2))^2;
        e12    := c12*e1*e2;
        f      := 1     +    e1 +       e2 +                  e12;
        fx     :=            c1*e1 +   c2*e2 +    (c1+c2)*e12;
        fxx    :=          c1^2*e1 + c2^2*e2 + (c1+c2)^2*e12;
        3*(fxx/f - (fx/f)^2):
      end_proc:

plot3d(Axes = Box, Ticks = 0,
    [Mode = Surface,
         [u, v, hold(30*kdv(u, 2*v, 3/5, 2/5))],
         u = [-30, 30], v = [-25, 25],
         Grid = [40, 40], Smoothness = [1, 0],
         Style = [HiddenLine, VLine]
    ]);
```

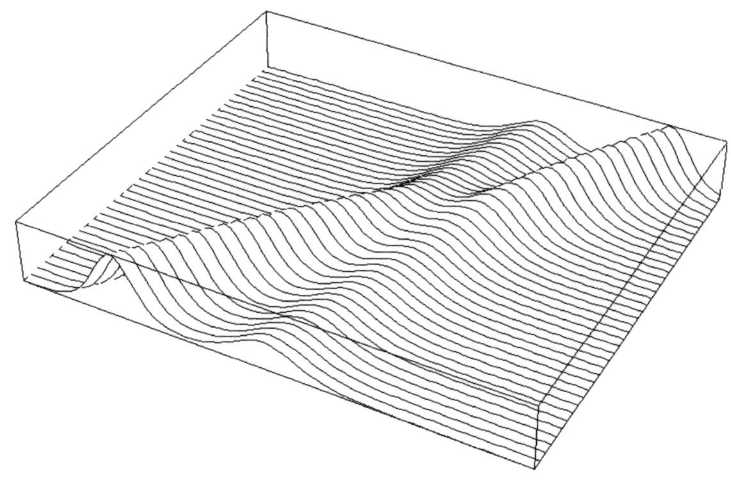

```
>> epiz_x := proc(a, b, u)
   begin
     (a+b)*cos(u) - a*cos((a+b)*u/a):
   end_proc:

epiz_y := proc(a, b, u)
   begin
     (a+b)*sin(u) + a*sin((a+b)*u/a):
   end_proc:

plot3d(Axes = Box, Ticks = 0,
       [Mode = Surface,
             [sin(v)*hold(epiz_x(1, 3, u)),
              sin(v)*hold(epiz_y(1, 3, u)),
              2*v],
             u = [0, 2*PI], v = [0, PI], Grid = [50, 20],
             Smoothness = [1, 1],
             Style = [HiddenLine, Mesh]
       ]);
```

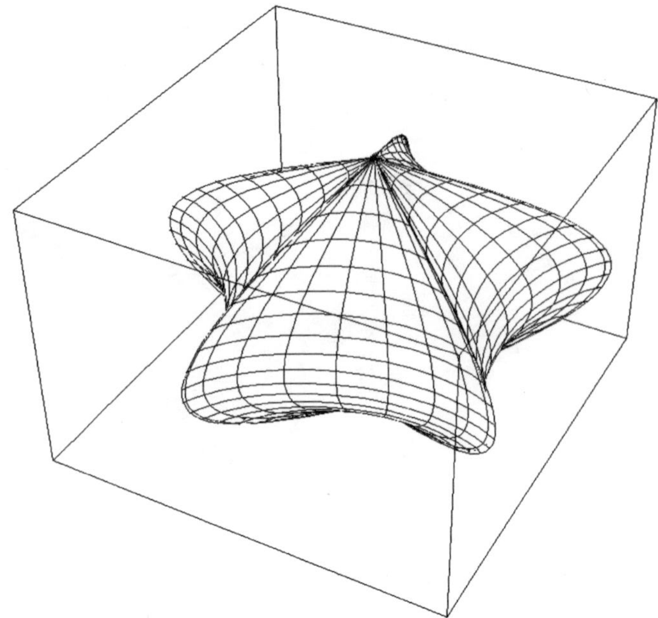

There is one point the user will have discovered, namely the fact that in order to prevent premature evaluation in the plot command, functions passed to the plot command are surrounded by a `hold`. While this is not really necessary for most of the examples in this section, it is very important for getting the correct plots for the imaginary part of the arcsine (see page 80) and for the real part of the arcsine (see page 81).

Of course, for three-dimensional objects we have to make the specification more detailed. Let us have a closer look:

```
>> epiz_x := proc(a, b, u)
   begin
     (a+b)*cos(u) - a*cos((a+b)*u/a):
   end_proc:

   epiz_y := proc(a, b, u)
   begin
     (a+b)*sin(u) + a*sin((a+b)*u/a):
   end_proc:

   plot3d(Axes = Box, Ticks = 0,
       [Mode = Surface,
           [v/3*hold(epiz_x(1, 3, u)),
            v/3*hold(epiz_y(1, 3, u)),
```

```
            v],
         u = [0, 2*PI], v = [-3, 3], Grid = [50, 20],
         Style = [HiddenLine, Mesh]
      ]);
```

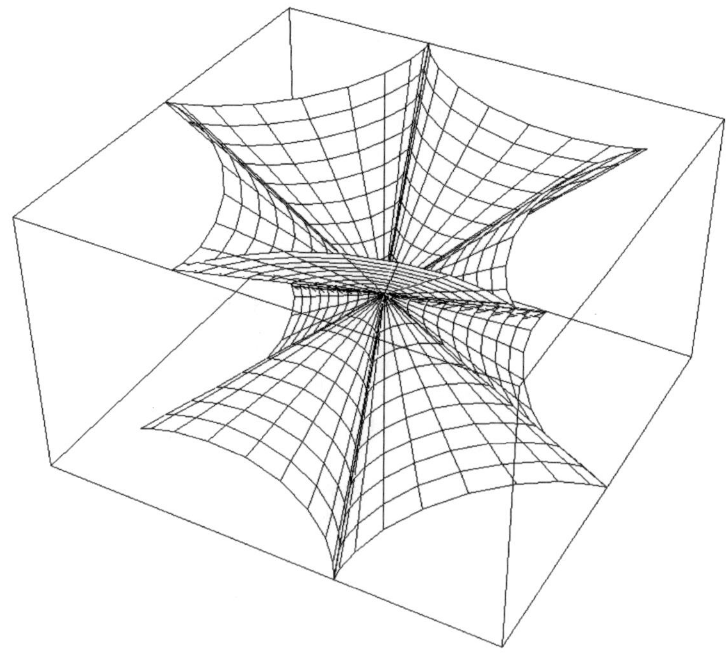

The user has to specify the **Mode**, which now allows the choice between more options than in the two-dimensional case. Furthermore, additional style options are available, and **Smoothness** and **Grid** can now be distinguished for the two different directions in the parameter space. Here we have plotted a surface **Mode = Surface** with the style option **Style = [HiddenLine, Mesh]**. By **Axes = Box** the axes were specified in such a way that the graphic is enclosed in a neat box.

Using the additional option **Mesh**, we required that both parameter lines had to be drawn. The different **Mesh** styles are illustrated by three spheres:

```
>> sphere_x := proc(u, v)
   begin
      sin(u)*cos(v):
   end_proc:

   sphere_y := proc(u, v)
   begin
```

```
      sin(u)*sin(v):
end_proc:

sphere_z := proc(u, v)
begin
  cos(u):
end_proc:

plot3d(Axes = Box, Ticks = 0,
Title = "Different Mesh Styles",
       [Mode = Surface,
            [3*hold(sphere_x(u, v)),
             3*hold(sphere_y(u, v)),
             3*hold(sphere_z(u, v))],
            u = [PI/2, PI], v = [-PI, PI],
            Grid = [20, 30], Style = [HiddenLine, Mesh]],
       [Mode = Surface,
            [2*hold(sphere_x(u, v)),
             2*hold(sphere_y(u, v)),
             2*hold(sphere_z(u, v))],
            u = [0, PI], v = [0, PI],
            Grid = [20, 20], Style = [HiddenLine, VLine]],
       [Mode = Surface,
            [hold(sphere_x(u, v)),
             hold(sphere_y(u, v)),
             hold(sphere_z(u, v))],
            u = [0, PI], v = [-PI, PI],
            Grid = [20, 20], Style = [HiddenLine, ULine]
       ]);
```

Different Mesh Styles

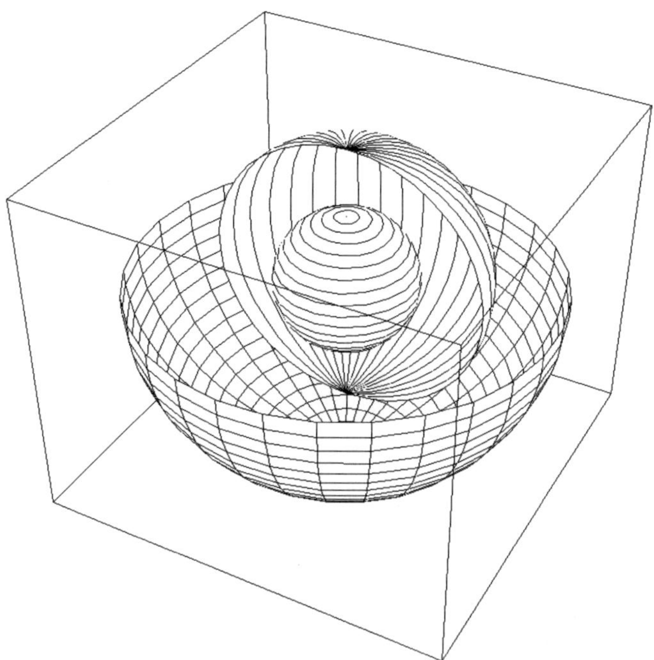

In three dimensions we can also draw `Curves` instead of `Surfaces`. For example the following curve plotted with the option `Style = [Impulses]`, a style which is also available for two-dimensional plots.

```
>> plot3d(Axes = Box, Ticks = 0,
          [Mode = Curve,
              [u, u, sin(u)],
              u = [-3.0, 3.0], Grid = [50],
              Smoothness = [2], Style = [Impulses]
          ]);
```

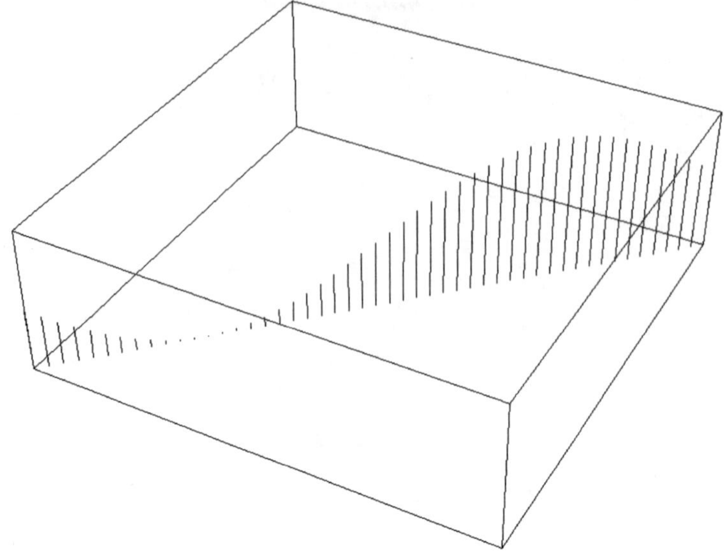

The different styles **Points**, **Lines**, **LinesPoints** and **Impulses** for curves are all combined in the following single plot:

```
>> plot3d(Axes = Box, Ticks = 0,
          Title = "Different Curve Styles",
          [Mode = Curve,
                  [u, -PI, sin(u)], u = [-PI, PI],
                  Grid = [50], Style = [Points]],
          [Mode = Curve,
                  [u, -PI/3, sin(u)], u = [-PI, PI],
                  Grid = [50], Style = [Lines]],
          [Mode = Curve,
                  [u, PI/3, sin(u)], u = [-PI, PI],
                  Grid = [50], Style = [LinesPoints]],
          [Mode = Curve,
                  [u, PI, sin(u)], u = [-PI, PI],
                  Grid = [50], Style = [Impulses]
          ]);
```

Different Curve Styles

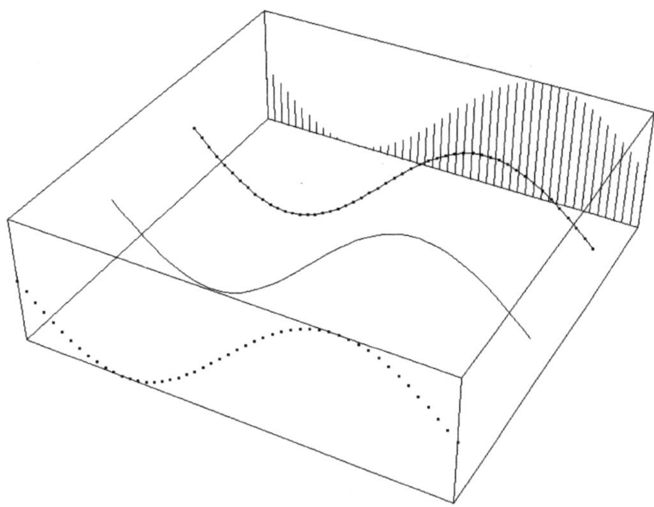

If no styles are specified then all parameter lines are drawn and a `WireFrame` representation is used instead of `HiddenLine`:

```
>> plot3d(Axes = Corner, Ticks = 0,
          [Mode = Surface,
                [u, v, 0.5*sin(u^2 + v^2)],
                u = [0, PI], v = [0, PI],
                Grid = [30, 30]
          ]);
```

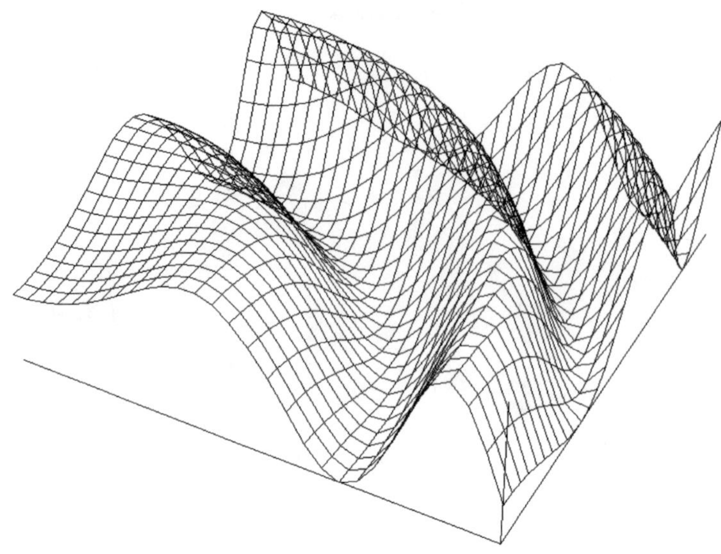

In three dimensions parametric surface plots are also possible:

```
>> plot3d(Axes = None, Scaling = UnConstrained,
        [Mode = Surface,
              [u*cos(v)*sin(u), u*cos(u)*cos(v), -u*sin(v)],
              u = [0, 3*PI], v = [0, PI],
              Grid = [20, 20], Smoothness = [2, 0],
              Style = [HiddenLine, Mesh]
        ]);
```

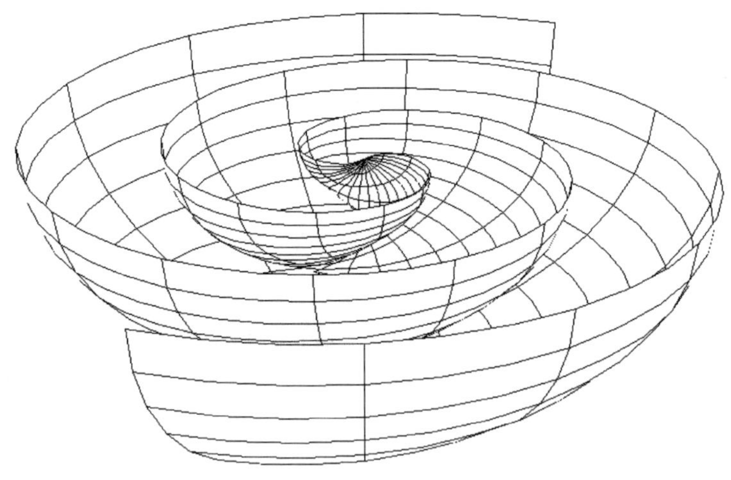

In the preceding plot the surrounding axes were prevented by `Axes = None` and an unrestrained scaling was allowed, which means that a scaling can be performed to make optimal use of the drawing window.

Of course, the `HiddenLine` style is also available for parametric surface plots:

```
>> plot3d(Axes = None,
          [Mode = Surface,
                 [(1.2)^v*sin(u)^2*cos(v),
                  (1.2)^v*sin(u)*cos(u),
                  (1.2)^v*sin(u)^2*sin(v)],
                 u = [0, PI], v = [-1, 2*PI],
                 Grid = [30, 30],
                 Style = [HiddenLine, Mesh]
          ]);
```

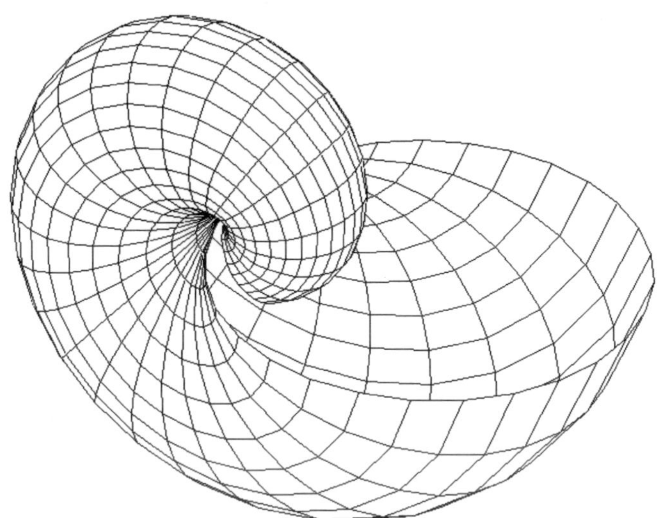

When no specification for the axes is made then ticked three dimensional coordinate axes are automatically drawn in one corner of the plot:

```
>> plot3d([Mode = Surface,
                 [u, v, sin(u*v)],
                 u = [-PI, PI], v = [-PI, PI],
                 Grid = [35, 35],
                 Style = [HiddenLine, Mesh]
          ]);
```

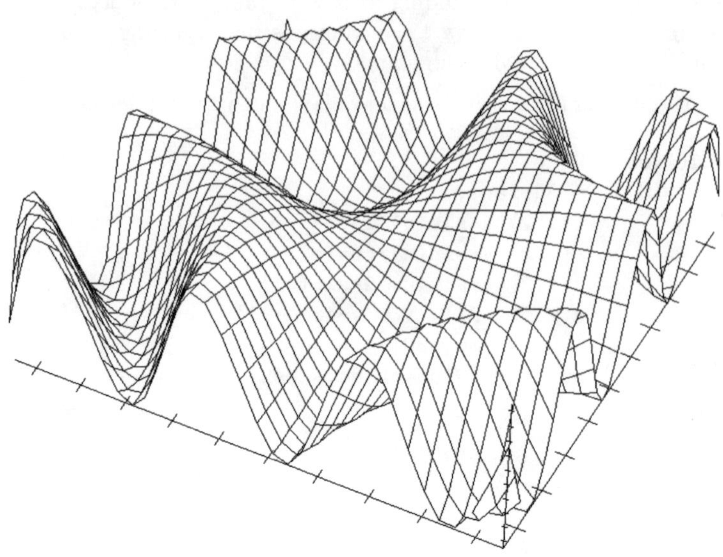

We conclude this section with some further examples, so that the reader can study
the effect of the different MuPAD options available in three dimensional plotting
for himself.

Parametric surface plot of the sine of the radius:

```
>> x := proc(u, v)
   begin
     v*sin(u):
   end_proc:

y := proc(u, v)
   begin
     v*cos(u):
   end_proc:

z := proc(u, v)
   begin
     1.5*sin(v):
   end_proc:

plot3d(Scaling = UnConstrained,
       [Mode = Surface,
          [hold(x(u, v)), hold(y(u, v)), hold(z(u, v))],
          u = [0, 2*PI], v = [0, 3*PI], Grid = [40, 40]
       ]);
```

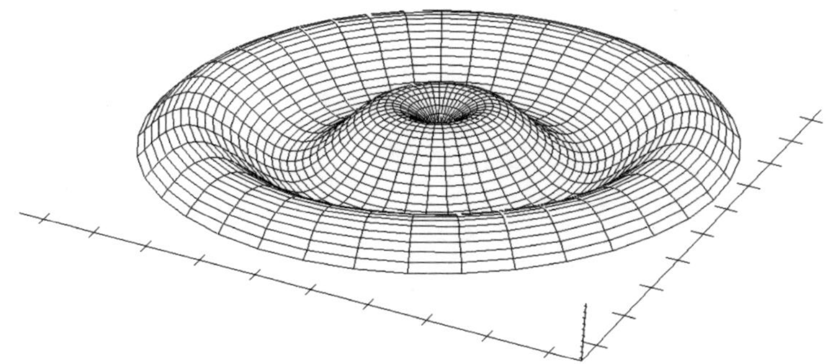

Four different surfaces combined in one plot:

```
>> plot3d(Axes = Box, Ticks = 0,
          CameraPoint = [-3.6, -4.9, 2.6],
          [Mode = Surface,
               [u, v, u^2], u = [-1, 1], v = [-1, -0.5],
               Grid = [30, 2], Style = [HiddenLine, Mesh]
          ],
          [Mode = Surface,
               [u, v, u^3], u = [-1, 1], v = [-0.5, 0],
               Grid = [30, 2], Style = [HiddenLine, Mesh]
          ],
          [Mode = Surface,
               [u, v, u^4], u = [-1, 1], v = [0, 0.5],
               Grid = [30, 2], Style = [HiddenLine, Mesh]
          ],
          [Mode = Surface,
               [u, v, u^5], u = [-1, 1], v = [0.5, 1],
               Grid = [30, 2], Style = [HiddenLine, Mesh]
          ]);
```

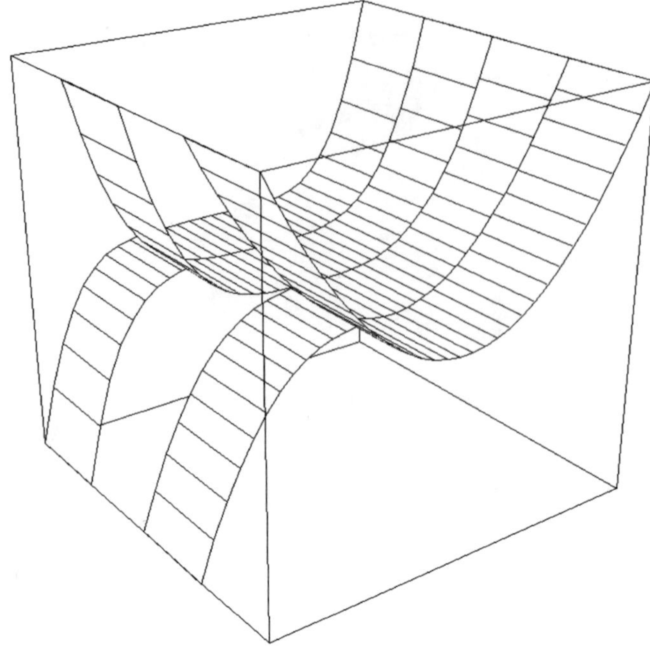

The imaginary part of the arcsine:

```
>> imaginary := proc(c)
   begin
     if domtype(c) =  DOM_COMPLEX
     then
       op(c, 2):
     else
       0:
     end_if:
   end_proc:

   plot3d(Axes = None, Scaling = UnConstrained,
          [Mode = Surface,
                  [u, v, hold(imaginary(asin(u+v*I)))],
                  u = [-4, 4], v = [-4, 4],
                  Grid = [30, 30],
                  Style = [HiddenLine, Mesh]
          ]);
```

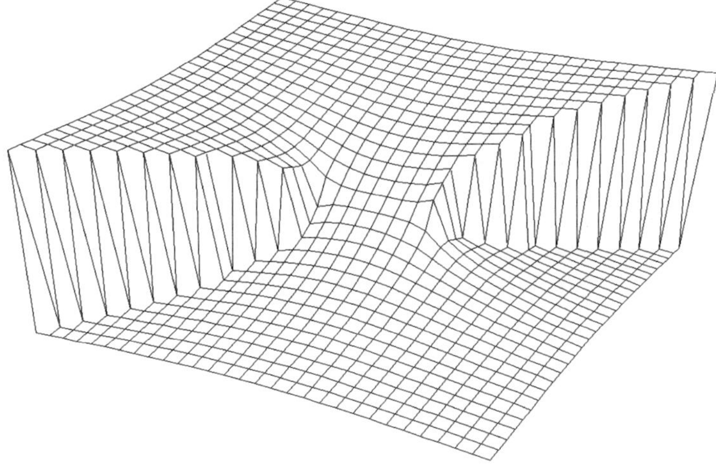

The real part of the arcsine:

```
>> real := proc(c)
   begin
     if domtype(c) = DOM_COMPLEX
     then
       op(c, 1):
     else
       0:
     end_if:
   end_proc:

   plot3d(Axes = None, Scaling = UnConstrained,
          [Mode = Surface,
                 [u, v, hold(real(asin(u+v*I)))],
                 u = [-4, 4], v = [-4, 4],
                 Grid = [30, 30],
                 Style = [HiddenLine, Mesh]
          ]);
```

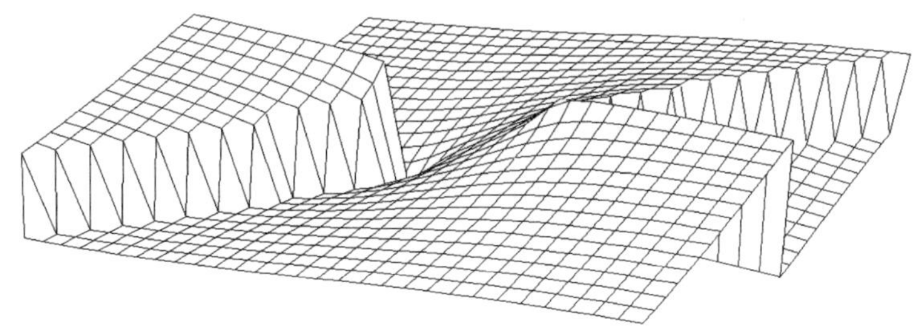

Plot of a minimal surface:

```
>> plot3d(Axes = Box, Ticks = 0,
          [Mode = Surface,
                [(3*u+3*u*v^2-u^3),
                 (3*v+3*u^2*v-v^3),
                 3*u^2-3*v^2],
                u = [-1.5, 1.5], v = [-1.5, 1.5],
                Grid = [30, 30],
                Style = [HiddenLine, Mesh]
          ]);
```

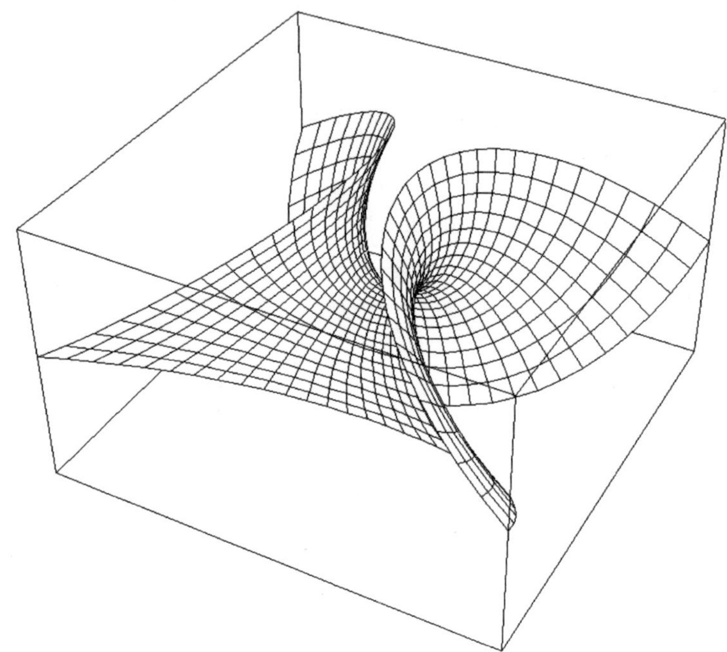

The difference made by smoothness:

```
>> plot3d(Axes = Box, Ticks = 0,
          CameraPoint = [0, 10, 4],
          [Mode = Surface,
                [1.2+sin(u)*cos(v), sin(u)*sin(v), cos(u)],
                u = [0, PI], v = [-PI, PI],
                Grid = [20, 15],
                Style = [HiddenLine, Mesh]
```

```
      ],
      [Mode = Surface,
             [-1.2+sin(u)*cos(v), sin(u)*sin(v), cos(u)],
             u = [0, PI], v = [-PI, PI],
             Grid = [20, 15], Smoothness = [0, 2],
             Style = [HiddenLine, Mesh]
      ]);
```

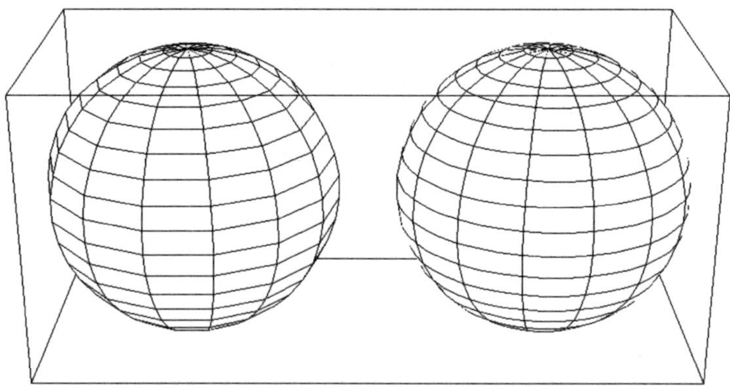

The wavy iterated sine:

```
>> plot3d(Axes = Box, Ticks = 0,
          [Mode = Surface,
                 [u, v, sin(u + sin(v))],
                 u = [-PI, PI], v = [-PI, PI],
                 Grid = [30, 30],
                 Style = [HiddenLine, Mesh]
          ]);
```

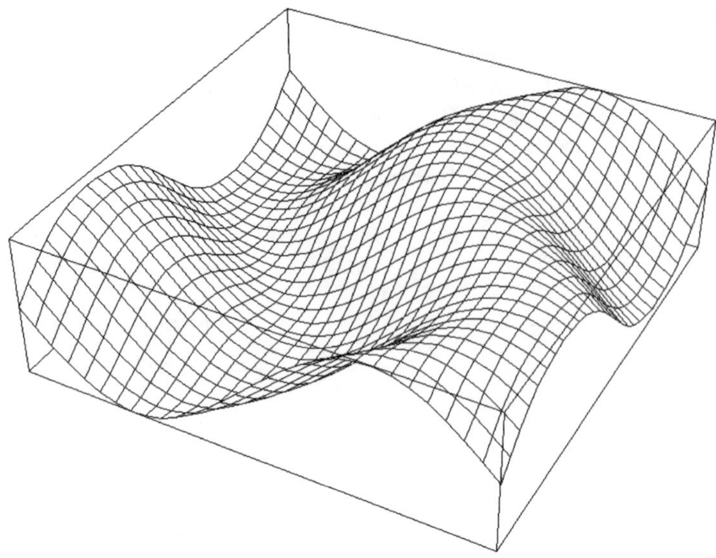

2.14   Errors and Debugger

In the interactive mode the most common error is that MuPAD is in a state where
the system apparently does not react to input. The reason for this is that a bracket
which has been opened has not yet been closed:

```
>> reset():

>> (1;

>>   2;

>> "What shall I do, MuPAD is not responding?";

>> "Try with input of a bracket";

>> "Now a sine-function and then a bracket";

>> sin(3.5);

>> );
-0.3507832276
```

Here the input of one or more right hand brackets helps to leave this confusing
state. A similar phenomenon occurs if a procedure has been opened but not closed
with **end_proc**:

```
>> F := proc() local i; option remember; begin procname(args());

>> 3;

>> 4;

>> end_proc;
proc()
  name F;
  local i;
  option remember;
begin
  procname(args());
  3;
  4
end_proc
```

The elimination of errors in user defined procedures is more difficult. For this purpose, MuPAD is provided with a convenient debugger. The appearance of the window based debugger interface depends on the operating system of the chosen hardware. However, the basic routines of the debugger are realized in the MuPAD kernel so, it is not surprising that there also is a pure terminal version of the debugger. Here is a brief introduction to this terminal version.

In the first example we shall ascertain a statement which generates a runtime error. The procedure `minus_list` describes an operation for lists which is analog to performing the difference operation for sets. For better reference we have numbered the lines of this program:

```
>> minus_list := proc(list1, list2)            #  1 #
                  local i, j;                   #  2 #
                  begin                         #  3 #
                    for i from 1 to nops(list1) do   #  4 #
                      for j from 1 to nops(list2) do  #  5 #
                        if list1[i] = list2[j] then   #  6 #
                          list1[i] := NIL;            #  7 #
                        end_if                        #  8 #
                      end_for                         #  9 #
                    end_for;                          # 10 #
                    list1                             # 11 #
                  end_proc:                           # 12 #

>> minus_list([a,c,b], [b,a]);
Error: Invalid index [list]
```

To find the error we start the **debug version** of MuPAD by calling either `mupad` **-g** on UNIX level or by the corresponding choice in the debug menu on the Macintosh. Both calls result in a standard MuPAD window to be opened. The only difference is that now, when reading files, additional information is generated. MuPAD not only memorizes which file contains which program, but also the line numbers of the corresponding program lines, since the debugger works line oriented. So, in order to use the debugger, procedures should be saved in line format. Now the file "tutorial" which contains the erroneous program should be read in:

```
>> read("tutorial"):
```

By using the **debug** command the error shall be detected:

```
>> debug(minus_list([a,c,b],[b,a])):
Enter procedure <minus_list>.
        args = [a, c, b], [b, a],
        proc depth = 1

Stop at line <4> in file <tutorial>.
```

MuPAD has now changed to the debug mode which is marked by the new prompt symbol mdx>. By calling c, the abbreviation for **continue**, we can run through the program until the error occurs:

```
mdx>c

Error: Invalid index [list]

Stop at line <6> in file <tutorial>.
```

In line 6, during the evaluation of a Boolean expression, a runtime error is generated. The reason for this is, that here we performed an invalid access to one of the lists. By using the command p (print) we can monitor the values of the indices i and j:

```
mdx>p i, j

i,j = 2, 2

mdx>p list1, list2

list1, list2 = [ c ], [ b, a ]
```

Thus we see, that calling `list1[3]` generated the runtime error. The question, however, still remains as to why the index i in the for-loop has been set to the value 3 at all, because `nops(list1)` should have given 1:

```
mdx>p nops(list1)

nops(list1) = 1
```

In order to monitor the steps of the evaluation we leave the procedure execution with q (quit) and repeat the program execution. Now we can follow its execution step-by-step:

```
mdx>q
```

```
>> debug(minus_list([a,c,b],[b,a])):
Enter procedure <minus_list>.
        args = [a, c, b], [b, a],
        proc depth = 1

Stop at line <4> in file <tutorial>.

mdx>s
Stop at line <5> in file <tutorial>.

mdx>s
Stop at line <6> in file <tutorial>.

mdx>s
Stop at line <6> in file <tutorial>.

mdx>s
Stop at line <7> in file <tutorial>.

mdx>s
Stop at line <5> in file <tutorial>.

mdx>s
Stop at line <6> in file <tutorial>.

mdx>s
Stop at line <7> in file <tutorial>.

mdx>s

Stop at line <6> in file <tutorial>.

mdx>s

Error: Invalid index [list]

Stop at line <6> in file <tutorial>.

mdx>
```

On taking a better look at the numbered lines the user can see that line 4, i.e.
the head of the external for-loop, is only executed once, whereas the body of the
loop was executed two times. This corresponds to the internal evaluation of the
for-loop, and it is precisely here that the error in the program minus_list is to

be found. The head, in particular `nops(list1)`, is only evaluated once. Thus, when the value of `list1` or the number of elements in `list1`, is changed, this has no influence on the range of the loop. This range remains unchanged. But, since in the above example `list1` became smaller whenever an element of `list1` was contained in `list2`, an error with respect to the index was the inevitable consequence.

The correct program should read as follows:

```
>> minus_list := proc(list1, list2)
              local i, j, offset;
              begin
                offset := 0;
                for i from 1 to nops(list1) do
                  for j from 1 to nops(list2) do
                    if list1[i-offset] = list2[j] then
                      list1[i-offset] := NIL;
                      offset := offset + 1;
                      break
                    end_if
                  end_for
                end_for;
                list1
              end_proc:
```

In the second example other debugger commands shall be illustrated. We will consider the procedure:

```
>> mistake:= proc(n)                        #112#
            local a, b, c;                  #113#
            begin                           #114#
              n;                            #115#
              nops_own(sin(x));             #116#
              c:=(12*n;float(PI)*n;144);    #117#
              a:=last(2);                   #118#
              b:=last(3);                   #119#
              a*b                           #120#
            end_proc:                       #121#
```

In this procedure the argument is evaluated and the number of operands, including the zero operand, of `sin(x)` is determined by the user defined procedure `nops_own`. After that, a rather senseless statement is executed. After accessing the second-last operand twice, i.e. the result of the call `nops_own(sin(x))`, the square of this operand is calculated. Therefore the result of `mistake`, independent of the parameter `n`, will always be 4, and the values of the variables `a` and `b` will always be 2. Surprisingly, we obtain:

```
>> mistake(12);
20736
```

Therefore, there must be a mistake somewhere which can be found by use of the
debugger. For this purpose the debug version of MuPAD is started again and,
to locate the error, the file "tutorial" is read in which contains the erroneous
program:

```
>> read("tutorial"):
>> debug(mistake(12));
Enter procedure <mistake>.
        args = 12,
        proc depth = 1

Stop at line <115> in file <tutorial>.

mdx> n

Stop at line <116> in file <tutorial>.

mdx> n

Stop at line <117> in file <tutorial>.

mdx> n

Stop at line <118> in file <tutorial>.

mdx> q

>>
```

The debugger yields information about the corresponding position within the file
under consideration. On input of **n** (abbreviation for **next**) MuPAD goes to the
next line. The debug mode can be left at any time by use of **q** (**quit**). This returns
us to the usual MuPAD surface. The command **n** always goes to the next line of
the program under investigation. The line structure of other procedures called
from the procedure under consideration, in this case **nops_own**, does not enter this
line count. If the user wants to work step-by-step through these other programs
then **s** (**step**) should be used instead:

```
>> debug(mistake(12));
.

.

.


Stop at line <115> in file <tutorial>.
```

```
mdx> s

Stop at line <116> in file <tutorial>.

mdx> s

Enter procedure <nops_own>
        from line <116> in file <tutorial>.
        args = sin( x ),
        proc depth = 2

Stop at line <47> in file <tutorial>.

mdx> q
>>
```

In this case information about user defined procedures accessed from the procedure under consideration is available.

However, it is rather boring to run step-by-step through procedures in big program packages. In this case breakpoints are preferred.

```
>> debug(mistake(12));
Enter procedure <mistake>.
        args = 12,
        proc depth = 1

Stop at line <115> in file <tutorial>.

mdx> S "tutorial" 118

mdx>c

Stop at line <118> in file <tutorial>.

mdx>c

Execution completed.

20736
```

With c (`continue`) the user immediately proceeds to the next breakpoint, here to line 118 which was created by the command S `"tutorial"` 118.

In this example running step-by-step through the procedure has not yet resulted in useful information about the cause of the error. To obtain such information the user needs to monitor the variables. This is done by the command D (`Display`), followed by the identifiers to be monitored. Monitoring of global, as well as local, variables, is possible:

```
>> debug(mistake(12));
Enter procedure <mistake>.
        args = 12,
        proc depth = 1

Stop at line <115> in file <tutorial>.

mdx> c

Stop at line <118> in file <tutorial>.

mdx> D a c

a = a
c = 144

mdx> n

Stop at line <119> in file <tutorial>.
a = 144
c = 144

mdx> n

Stop at line <120> in file <tutorial>.
a = 144
c = 144

mdx> n

Stop at line <121> in file <tutorial>.
a = 144
c = 144

mdx> n

Execution completed.

20736
```

Here, the user can see that the display command results in the printing of the variables under consideration at each Stop.

In the meantime the reader must certainly have noticed that there is a mistake at
the position of the last-assignment to a. Therefore, we should test to see if the
program works correctly if a attains its correct value. To try this we need to change
the value of a interactively during a debug run, so that the remaining statements
are executed with the corrected value of a. To perform this value change we have
the Execute-command e:

```
>> debug(mistake(12));
Enter procedure <mistake>.
        args = 12,
        proc depth = 1

Stop at line <115> in file <tutorial>.

mdx> c

Stop at line <118> in file <tutorial>.

mdx> n

Stop at line <119> in file <tutorial>.

mdx> e a:=12; b:=12;

mdx>p a, b
a, b = 12, 12

mdx>c

Execution completed.

1728
```

In this case the value 12 was assigned to a as well as to b and then the computation
was continued.

Unfortunately even this correction did not have the desired result. So something
must be wrong with the assignment to b. In order to correct that, we make some
improvements in the file "tutorial" and read in this file anew. This results in
warnings which indicate that procedures and MuPAD data were overwritten:

```
>> read("tutorial"):
Warning: Overwrite definition of procedure <minus_list>,
        line <1>, file <tutorial>.
New definition: line <1>, file <tutorial>
Warning: Overwrite definition of procedure <fct>,
        line <25>, file <tutorial>.
```

```
New definition: line <25>, file <tutorial>
Warning: Overwrite definition of procedure <FUNC>,
        line <15>, file <tutorial>.
New definition: line <15>, file <tutorial>
Warning: Overwrite definition of procedure <nops_own>,
        line <45>, file <tutorial>.
New definition: line <45>, file <tutorial>
```

.

.

.

If the user wants to get rid of breakpoints then the command C (**Clear**) should be used:

```
>> debug(mistake(12));
Enter procedure <mistake>.
        args = 12,
        proc depth = 1

Stop at line <115> in file <tutorial>.

mdx> C "tutorial" 118

mdx> c

Execution completed.

20736
```

The next **continue** disregards this breakpoint and completes the execution of the program without stopping.

With respect to the program error we have not yet found out why the assignments did not have the desired effect. The reason for this is that we have not yet completely understood the interplay between statement sequences and the function **last**.

In order to understand the error in the procedure **mistake** the user should note that the value of an expression sequence, consisting of statements, is indeed the value of the last statement:

```
>> bool((2; 4; 7)=7);
TRUE
```

But, nevertheless, the function **last** counts all statements which were executed in this sequence:

```
>> (2; 3; 4);
4
```

```
>> print(%, %2, %3);
4, 3, 2
```

It is not always worthwhile to work with the debugger. If a procedure only consists of a few lines then it is easily analyzed by increasing the depth of output. In this case intermediate results are also printed:

```
>> PRINTLEVEL := 40:
```

```
>> mistake(12);
               12
                2
                2
              144
         3.769911184e1
              144
             c:=144
             a:=144
             b:=144
             20736
20736
```

Here the user sees that something must be wrong with the assignments to **a** and **b**.

Users of MuPAD versions 1.0 or 1.1 should observe that there is a systematic bug in the debugger. There, the generation of additional information for procedures leads to the creation of new operands. Thus, in these MuPAD versions, the debugger should not be used to debug procedures performing program manipulation. This bug has been abolished in subsequent versions.

2.15 The Profiler

In order to analyse the runtime behavior of a program in detail it is necessary to protocol the runtimes of the procedures it calls. The number of procedure calls gives further information about the program behavior. With the help of this information, time-critical or runtime-intensive parts of programs can be identified. Now the user can easily detect which parts of the program need to be optimized in order to shorten the total runtime. The testing of the efficiency of the option remember in a procedure, which is often used for runtime optimization, is a further feature of the runtime analysis of a program.

By using the command **profile(statement)** the runtime protocol of a program **statement** is shown on the monitor. The profile of the following program, which calculates the n-th Fibonacci number, shows the structure of a run time protocol:

```
>> reset():
```

```
>> fib1:=proc(n)
begin
   if n<2 then n else fib1(n-1)+fib1(n-2) end_if:
end_proc:

>> profile(fib1(10));

Total time: 200 ms
------------------
fib1: 100 % 200 ms total 177 call(s) 0 lookup(s) 1.1 ms/call

<fib1> calls
   fib1 : 176 time(s)
```

The calculation of the 10th Fibonacci number takes 200ms. As in this example the program consists of only one procedure `fib1` the total runtime is also the runtime of the procedure. The runtime of `fib1` is therefore 100% of the total runtime. The procedure `fib1` was called 177 times. The mean runtime (arithmetical mean) per call of `fib1` is 1.1ms. The second part of the protocol shows for each called procedure $< proc >$ the procedures which were directly called in $< proc >$. In additon the amount of calls is noted. The output above shows that the procedure `fib1` calls itself recursively.

```
>> fib2:=proc(n)
   option remember;
begin
   if n<2 then n else fib2(n-1)+fib2(n-2) end_if:
end_proc:

>> profile(fib2(10));

Total time: 20 ms
------------------
fib2: 100 % 20 ms total 19 call(s) 8 lookup(s) 1.0 ms/call

<fib2> calls
   fib2 : 18 time(s)
```

In contrast to the procedure `fib1` `fib2` uses the option remember. Now only 19 calls of `fib2` are necessary to calculate the 10th Fibonacci number. Eight of these calls calculated their solutions by looking them up in the remember table and not by executing the procedure body.

To illustrate this further here is a more complicated example. The matching of the bipartite graph which consists of the node set $V \cup U$ and the edge set EDGE is to be calculated. The corresponding program can be found in section 3.2.5.2.

```
>> V:={v1,v2,v3,v4,v5,v6}:
U:={u1,u2,u3,u4,u5,u6}:

EDGE:={[v1,u1], [v1,u2], [v1,u4], [v2,u2], [v2,u6], [v3,u2],
       [v3,u3], [v4,u3], [v4,u5], [v4,u6], [v5,u3], [v5,u4],
       [v5,u5], [v5,u6], [v6,u2], [v6,u5]}:

>> profile(bipartite_matching(V, U, EDGE));


Total time: 340 ms
------------------
build_A             : 47.0 % 160 ms total 7 call(s) 0 lookup(s)
                              22.8 ms/call
loop                : 20.5 %  70 ms total 7 call(s) 0 lookup(s)
                              10.0 ms/call
build_Q             : 14.7 %  50 ms total 7 call(s) 0 lookup(s)
                               7.1 ms/call
augment             :  8.8 %  30 ms total 8 call(s) 0 lookup(s)
                               3.7 ms/call
bipartite_matching:   5.8 %  20 ms total 1 call(s) 0 lookup(s)
                              20.0 ms/call
init                :  2.9 %  10 ms total 7 call(s) 0 lookup(s)
                               1.4 ms/call

<loop> calls
   init    : 6 time(s)
   augment : 6 time(s)

<bipartite_matching> calls
   init : 1 time(s)

<augment> calls
   augment : 2 time(s)

<init> calls
   loop    : 7 time(s)
   build_Q : 7 time(s)
   build_A : 7 time(s)
```

Only user defined procedures, not system functions are protocolled. If system functions need to be protocolled a runtime protocol can be obtained by simply re-defining system functions as procedures.

This shall be illustrated on the heuristic algorithm for calculating the greatest common divisor of two simple polynomials with integer coefficients. The corresponding program can be found in section 3.2.2.1.

```
>> a := poly(
   9*x^5 + 2*x^4*y*z - 189*x^3*y^3*z + 117*x^3*y*z^2 + 3*x^3 -
   42*x^2*y^4*z^2 + 26*x^2*y^2*z^3 + 18*x^2 - 63*x*y^3*z +
   39*x*y*z^2 + 4*x*y*z + 6,
   [x, y, z]
):

b := poly(
   6*x^6 - 126*x^4*y^3*z + 78*x^4*y*z^2 + x^4*y + x^4*z + 13*x^3 -
   21*x^2*y^4*z - 21*x^2*y^3*z^2 + 13*x^2*y^2*z^2 + 13*x^2*y*z^3 -
   21*x*y^3*z + 13*x*y*z^2 + 2*x*y + 2*x*z + 2,
   [x,y, z]
):

>> profile(heu_gcd(a, b));

Total time: 140 ms
-------------------
heu_gcd: 99.9 % 140 ms total 3 call(s) 0 lookup(s) 46.6 ms/call

<heu_gcd> calls
   heu_gcd : 2 time(s)
```

To examine the runtime in more detail the system functions igcd, genpoly and
evalp shall also be protocolled.

```
>> igcd_sys := igcd:
igcd := proc() begin igcd_sys(args()) end_proc:
genpoly_sys := genpoly:
genpoly := proc() begin genpoly_sys(args()) end_proc:
evalp_sys := evalp:
evalp := proc() begin evalp_sys (args()) end_proc:

>> profile(heu_gcd(a, b));

Total time: 150 ms
-------------------
heu_gcd: 53.3 % 80 ms total 3 call(s) 0 lookup(s) 26.6 ms/call
evalp  : 39.9 % 60 ms total 6 call(s) 0 lookup(s) 10.0 ms/call
genpoly:  6.6 % 10 ms total 3 call(s) 0 lookup(s) 3.3 ms/call
igcd   :  0.0 %  0 ms total 1 call(s) 0 lookup(s) 0.0 ms/call

<heu_gcd> calls
   genpoly : 3 time(s)
   evalp   : 4 time(s)
```

```
   igcd    : 1 time(s)
   heu_gcd : 2 time(s)

<igcd> calls
   evalp : 2 time(s)
```

One more comment about call graphs: At first sight the statement that the procedure `igcd` calls `evalp` seems to be incorrect. However, the evaluation of an argument procedure is considered as part of the procedure execution in MuPAD, therefore, the statement above is true, as `igcd(evalp(a,x=xv), evalp(b,x=xv))` is carried out.

2.16   Underline Functions

The MuPAD kernel contains a number of functions not usually accessed by a function call but rather by some kind of operational notation. For example `*`, `+`, `^`, or `break` or the `case`-statement, to name but a few. All these have a functional equivalent. For `+` we have already seen that the functional equivalent is `_plus`. This is also the case with all other operators. There are internal system functions which can be used instead of these. These system functions are the so-called underline functions. Their name is explained by the fact that they all start with an underline (`_`).

A list of existing underline functions is easily obtained by use of the function `get_ufunc` which can be found in the file `"tutorial"`:

```
>> reset():
```

```
>> get_ufunc();
   [_for_in_par, _repeat, _fconcat, _mod, _case, _seqgen, _if,
   _union, _assign, _div, _less, _procdef, _stmtseq, _unequal,
   _minus, _power, _for, _index, _equal, _while, _and, _for_par,
   _leequal, _next, _for_down, _concat, _break, _intersect,
   _parbegin, _mult, _quit, _plus, _not, _for_in, _exprseq,
   _seqbegin, _or, _range]
```

Among these underline functions, which are equivalents of the basic building blocks for programming, the user finds the `for`-loop, the commands `break` and `next` and the logic operators `and`, `or` etc.

The user has direct access to these underline functions, for example instead of `a+b` he can use `_plus(a, b)` or instead of `A := 12` the same result is achieved by `_assign(A, 12)`:

```
>> _plus(a, b);
   a+b
```

```
>> _assign(A, 12): A;
     12
```

Similarly, the user can access control structures. For example, in order to find the functional equivalent of the following `for`-statement

```
for a from 2 to 8 do print(a) end_for;
```

The best way to analyze this statement is by using `op`. To prevent evaluation a `hold` is necessary:

```
>> op(hold((for a from 2 to 8 do print(a) end_for)), 0);
     _for
```

```
>> op(hold((for a from 2 to 8 do print(a) end_for)));
     a, 2, 8, NIL, print(a)
```

The first output, `_for`, yields the name of the functional equivalent, and in the second output we find the operands to be inserted in this function, here `a, 2, 8, NIL, print(a)`. The `NIL` here represents the missing specification of the step width. If all these operands are inserted in the underline function `_for`, then this leads to the same result as the original `for`-statement:

```
>> _for(a, 2, 8, NIL, print(a));
     2
     3
     4
     5
     6
     7
     8
```

However, it is strongly recommended not to use these underline functions. These are only an additional tool for the very experienced MuPAD programmer. Contrary to other functions, when working with underline functions, MuPAD only performs a restricted semantic check of the operands: 

```
>> 12 := 24;
     Syntax Error: Unexpected symbol in assignment

     12 := 24;
        ^
```

```
>> _assign(12, 24);
     Error: Invalid left hand side
```

In the first case, by checking the syntax, MuPAD's parser detected that this cannot be a reasonable assignment. In the second case such a check by the parser did not take place, and the error was only detected when the assignment was carried out. In this case, no problems were caused. However, if we call **_assign(A)**, i.e. a call instead on an assign function with only one argument, a system breakdown is the result.

A further warning is necessary, because manipulation of the underline function can change the system beyond recognition:

```
>> _plus := _power:
```

```
>> 2+10;
       1024
```

MuPAD is however, still able to add numbers, but by the assignment **_plus :=** **_power** the operation **_power** has replaced the usual operation **_plus**, thus the result of 2+10 is 2 to the power 10. Although the redefinition of system functions with the underline function is dangerous, it is nevertheless a powerful tool for the experienced programmer. By the way, these redefinitions can also be performed locally:

```
>> reset():
```

```
>> plus_localnew := proc(a, b)
                 local _plus;
              begin
                _plus := _mult;
                a+b
              end_proc:
```

```
>> plus_localnew(12, 12);
       144
```

```
>> 12+12;
       24
```

One advantage in using underline functions is that these constitute efficient means for brief and concise programming. For example, the following program for the computation of the arithmetic sum of **count** elements with the starting point **start** and step width **step**:

```
>> arithm_series := proc(start, stepp, count)
                 local i;
              begin
                _plus(start+i*stepp $ i=0..(count-1))
              end_proc:
```

```
>> arithm_series(2, 3, 2);
      7
```

2.17 Functional Environments

MuPAD's programming language has a similar syntax to Pascal. However, all statements and expressions are represented functionally. Functions and their associated objects are stored in the form of this functional representation. Generally, a function is represented by an identifier or a so-called *functional environment*. If the representation is carried out by an identifier then, in nearly all cases, this has a functional environment assigned to it.

The data type *DOM_FUNC_ENV* is used to store the functional environments. The relevant data structure has three entries. The first two can be of any type. However, in most cases these are objects of the data type *DOM_EXEC, DOM_PROC* or *DOM_EXT*. The third entry is either empty or it contains a table.

In the use of the functional environments two situations can be distinguished from one another. If they appear in an expression (data type *DOM_EXPR*) as the zero operand then they determine the evaluation and the output of the functional expression. The evaluation is determined by the first operand. The second operand determines the output functionality. In all other cases the functional environments play no special roles, i.e. they are treated as usual objects. In the following we would like to demonstrate by some examples how functional environments can be used.

All MuPAD constructs are internally represented by functional environments. Therefore, these are themselves absolutely normal data which are available to the user. The experienced programmer has the opportunity of adapting the system to his requirements by selective manipulation of these functional environments. However, for the inexperienced MuPAD user it is urgently recommended not to manipulate the functional environments: A number of system breakdowns may be the consequence of careless manipulation. 

As the first example we shall look at one such functional environment:

```
>> op(_exprseq);
    _exprseq, _exprseq, table("type"="_exprseq")
```

The first two operands are the so-called *executables* (data type *DOM_EXEC*). The first one is responsible for the evaluation of the functional environment and consists of four operands.

```
>> op(op(_exprseq, 1));
    55, NIL, "_exprseq", NIL
```

The first is a natural number and is a reference to an internal C routine. The second operand is either empty *DOM_NIL* or a natural number which is also a reference to a C routine neccessary for evaluation. The third operand is a character string giving the name of the function and the fourth is either empty or contains a remember table for the function.

The second *DOM_EXEC* object also contains four operands and is responsible for output.

```
>> op(op(_exprseq, 2));
     51, 2, ", ", "_exprseq"
```

The first operand is once again a number which refers to an internal C routine. The second one is a number which determines the binding priority of the output. If this value is changed it influences the brackets in the output. The third operand contains a symbol for the output in operator style and the final operand contains a character string for the functional output.

If, for example, the user is uncomfortable with the fact that the elements of an expression sequence are seperated by a comma instead of the word `Seperator`, this can be changed at will:

```
>> _exprseq := func_env(built_in(55, NIL, "_exprseq", NIL),
                        built_in(51, 2, " Separator ", "_exprseq"),
                        NIL):
```

Firstly, a new functional environment is created by using the functions `func_env` and `built_in`. To do this we copied the neccessary entries from above and assigned them to `_exprseq`. Since internally the functional environment for expression sequences (the comma operator) is accessed via the identifier `_exprseq`, this new functional environment is automatically used in the following:

```
>> (a, b, c, (d, k));
     a Separator b Separator c Separator d Separator k
```

Due to the fact that the numbers of the internal C routines are difficult to remember, the user can achieve this effect in a more concise (and elegant) way by selective manipulation of the already present functional environment:

```
>> _exprseq := subsop(_exprseq, [2,3] = ", "):
```

By the selective manipulation of the operator symbol we have succeeded in restoring the former appearence of expression sequences.

```
>> (a, b, c, (d, k));
     a, b, c, d, k
```

Instead of changing one single entry in an existing *DOM_EXEC* element the user can replace the operand responsible for output. To demonstrate this we shall consider the absolute value function `abs`:

```
>> abs(2); abs(a);
     2
   abs(a)
```

Now we want to change the output so that a non-evaluable call `abs(a)`, e.g. a call
with a non-numerical argument a, displays the symbol |a| on the screen. As we
already know the second operand of the functional environment is responsible for
output, so we will have to modify it. For this purpose we shall write a function
that returns its argument enclosed in the absolute value sign:

```
>> f := proc(x)
       local h;
       begin
           h := expr2text(op(x,1));
           "| ".h." |";
       end_proc:
```

Now we can substitute the second operand in the functional environment of `abs`
by this procedure:

```
>> abs := subsop(abs, 2 = f):
```

The following example shows that these few commands have had the desired effect:

```
>> abs(1); abs(-2); abs(a); abs(x+y);
       1
       2
     | a |
   | x+y |
```

Only the output is affected by these changes. They have absolutely no influence
on the input. Consequently, the entry |a| leads to a syntax error:

```
>> |a|;
     Syntax Error: Unexpected character

     |a|;
      ^
```

Apart from changing the output functionality, the first operand (which is responsible for evaluation) can also be manipulated and replaced. In the following example
we would like to enhance to the functionality of the sine function.

```
>> sin(-4); sin(asin(x)); sin(atan(x));
       sin(- 4)

   sin(asin(x))

   sin(atan(x))
```

We want to equip the sine with some elementary simplification rules, like the
extraction of a negative sign in a whole number or rational argument, or the

elimination of the sine call and its reverse function. Naturally, we need to keep the old functionality and so first we must store this datum:

```
>> _sin := sin:
```

Now we shall write a procedure in which we implement the desired rules:

```
>> f := proc(x)
     local f;
begin
     if args(0) <> 1 then error("wrong no of args") end_if;

     case type(x)
     of DOM_INT do
     of DOM_RAT do
            if x < 0 then return(-_sin(-x)) end_if;
            break;

     of "asin" do
            if nops(x) <> 1 then break end_if;
            f:= op(x,1);
            return(f);

     of "acos" do
            if nops(x) <> 1 then break end_if;
            f:= op(x,1);
            return(sqrt(1-f^2));

     of "atan" do
            if nops(x) <> 1 then break end_if;
            f:= op(x,1);
            return(f/sqrt(1+f^2));
     end_case;

     _sin(x)
end_proc:
```

This procedure can of course be added to. Substitution of the operand responsible for evalution in the sine functional environment transfers these rules to the sine, so "the sine learns something new."

```
>> sin:= subsop(_sin, 1=f):
sin:= subsop(sin, [1,5]=op(_sin, [1,4])):
```

With the second statement, we make a copy of the remember table of the original sine.

```
>> sin(-4); sin(asin(x)); sin(atan(x));
        - sin(4)


           x


     x * (x^2 + 1)^(-1/2)
```

By using the functional environment of the original sine and retaining its name, these modifications are a real complement of the functionality. The other functions deal with this "new" sine exactly as they did before:

```
>> diff(sin(x^2), x);
     x * cos(x^2) * 2
```

We still have not explained the aim and purpose of the third operand in the functional environment. Here, the functional environment can be given attributes in a table. If the user does not like the output of the functional environment _plus:

```
>> reset():

>> _plus;
     built_in( 63, NIL, "_plus", NIL )
```

he can enter the desired form under the attribute "print":

```
>> _plus := funcattr(_plus, "print", "Addition"):

>> _plus;
     Addition
```

However, this entry has no influence on the expression output formed by _plus. It only affects the output of the functional environment provided that this is not the zero operand in an expression.

```
>> a+b; [a, _plus, b];
           a+b
     [a, Addition, b]
```

The differentiation routine in the MuPAD kernel knows the rules for differentiation and the derivative of certain elementary functions. If a function f, which cannot be differentiated by the differentiation routine using its included rules, is contained in an expression then the call $\text{diff}(f(x), x)$, for instance remains unevaluated. With the aid of the functional environment and the attribute table there is the possibility to append rules to the differentiation routine. This is carried out on the user level and not in the kernel. For this purpose the user can enter a procedure fdiff, which determines the derivative of f, under the index "diff" in the attribute table.

If the differentiation routine now finds an unknown function, it first checks in the attribute table for an entry in the index `"diff"`. If an entry exists then the procedure fdiff is called in the form fdiff($expr, var$). Here $expr$ is the expression to be differentiated and var is the differentiation variable. As an example we shall now look again at the absolute value function `abs`. This can be differentiated in $\mathbb{R}\backslash\{0\}$ and we have

$$\frac{d\,abs(x)}{d\,x} = \frac{abs(x)}{x}$$

```
>> diff(abs(x), x);
    diff(abs(x), x)
```

MuPAD does not yet know this rule, therefore, we enter it in the `abs` attribute table:

```
>> abs := funcattr(abs, "diff",
        proc(expr,var)
        local e1;
        begin
            e1 := op(expr,1);
            (expr/e1) * diff(e1, var)
        end_proc):
```

```
>> diff(abs(x),x);
    1/x * abs(x)
```

```
>> diff(abs(x^2),x);
    1/x * abs(x^2) * 2
```

This very useful characteristic can of course be used for the user's own procedures of the data type *DOM_PROC*. If `f` is a procedure then the user can embed it in a functional environment with the call:

```
funcattr(f, index, attr)
```

In this functional environment the datum `attr` is entered in the attribute table in the index `index`.

2.18   Program Manipulation

For the further development of MuPAD the fact that procedures can be manipulated by the system itself, in the same way as all other MuPAD data, will play an important role. This will lead to the development of procedures for optimization and parallelization of MuPAD code. Such procedures for optimization and parallelization will be written in MuPAD language itself. Up to now we have not provided these instruments. However, we want to exhibit the possibilities given by MuPAD's ability for program manipulation.

As a first example we shall consider the procedure `factorial` which computes the factorial of a given number **n**:

```
>> reset():
```

```
>> factorial := proc(n)
              local i, p;
           begin
             p := 1;
             for i from 2 to n do p := p*i end_for
           end_proc:
```

```
>> factorial(80);
    7156945704626380022948115337231865321655846573420\
    3657525771094450582270392554801488426689448672800\
    0814080000000000000000000000
```

By use of **subsop** it is easily possible to change this into a procedure which computes the sum of the numbers **1** to **n**, instead of their product:

```
>> sum := subsop(factorial, [4, 1, 2]=0, [4, 2, 2]=1,
              [4, 2, 5, 2]=p+i, 6 = sum);
    proc(n)
    name sum;
    local i, p;
    begin
      p := 0;
      for i from 1 to n do
        p := i+p
      end_for
    end_proc
```

```
>> sum(500);
    125250
```

The procedures sum and factorial differ in the start value for the **for**-loop `p :=
0;` as well as in the operation by which the numbers are combined; these were the operands with the path names `[4, 1, 2]` and `[4, 2, 5, 2]`, which were changed by the corresponding call on **subsop**.

As motivation for the next example the user should observe that in procedures local variables are often declared which, after a number of improvements in the program, are no longer necessary. These local variables should be erased since they only slow down the program. However, for long procedures it is not easy to

see which local variables are necessary and which are not. Therefore, we want to
write a function for optimizing procedures in the sense that it erases unnecessary
local variables automatically. We call this function `purge_locals`. Its code, also
contained in `"tutorial"`, is:

```
>> purge_locals :=  proc(f)
                  local dummy, a, i;
               begin
                 a := {};
                 for i in op(f, 2) do
                    if subs(op(f,4), i=dummy) <> op(f,4) then
                      a := {i, op(a)}
                    end_if
                 end_for;
                 subsop(f, 2=op(a))
              end_proc:
```

Here we simplified the procedure so that it is not able to erase unnecessary variables
with the name `dummy`. However, this restriction is easily abolished by the more
experienced programmer. One way to do this, is to generate a "new" identifier with
the MuPAD function `genident` by `genident(op(f,2),op(f,4))`, which gives an
identifier not contained in the locals and the body of the procedure. This identifier
serves as a dummy identifier for the necessary substitution. Now, we consider the
function,

```
>> function := proc(a)
                  local i, G, DIGITS, LEVEL;
               begin
                 i := a^2;
                 DIGITS := 20;
                 print(i)
               end_proc:
```

```
>> function(3.2);
     1.02400000000e1
```

which contains the unnecessary variables `G` and **LEVEL**. Indeed, these are easily
removed by our function `purge_locals`:

```
>> function := purge_locals(function): function;
     proc(a)
     name function;
     local i, DIGITS;
     begin
       i := a^2;
       DIGITS := 20;
```

```
    print(i)
  end_proc
```

Here, obviously, it was immaterial whether the local variables to be removed have a value outside the environment of the considered **function** or not.

In order to measure the efficiency of the **case**-statement we want to compare the system time necessary for a function containing 500 **case**-statements with that of the corresponding function containing 500 **if**-statements. To do this we consider programs of a type

```
>> CASE := proc ( a )
        begin
          case a
            of 0 do 0; break
            of 1 do 1; break
            of 2 do 2; break
            of 3 do 3; break
          end_case
        end_proc:
```

where the argument a, beginning with 0, is compared successively with a number increased by 1 in each step. Such a comparison can also be achieved by use of the logically equivalent **if**-statement:

```
>> IF := proc ( a )
       begin
         if a<=0 then
           0
         else
           if a<= 1 then
             1
           else
             if a<= 2 then
               2
             else
               AA
             end_if
           end_if
         end_if
       end_proc:
```

However, we want to have 500 comparisons instead of the three above. Writing 500 such loops will certainly be a boring task. However, this can be facilitated by using MuPAD's ability for program manipulation.

A corresponding program with 500 **case**-statements is easily generated, for example by:

```
>> pp := proc(a)  begin AAA end_proc:
```

```
>> FF(a, subs(hold((LL, (LL; break)))), LL=LL + k) $ k=0..500):
```

```
>> CASE := subs(pp, AAA=%):
```

```
>> CASE := subs(CASE, LL=0, FF=_case):
```

Here we assumed that the identifiers **AAA**, **LL**, **FF**, **a** and **pp** are not assigned. With the first command, we generated a procedure containing the identifier **AAA** as a place holder in its procedure body. After that a new procedure body was made available by calling on the unassigned procedure **FF**. FF contains the place holder **LL**, which was iterated 500 times so that for **LL** = 0 its arguments were those of the desired **case**-statement. Subsequently, this procedure call was inserted instead of the place holder. Then **LL** was equated to 0 and the formal function **FF** was replaced by the underline function **_case**.

Trying this generation for 3, instead of 500, results exactly in the procedure **CASE** which was given before. The corresponding **IF** program is generated similarly:

```
>> LL := NIL: FF := NIL:
```

```
>> IF := proc(a) begin FF(a<=LL, LL, AA) end_proc:
```

```
>> (IF := subs(IF, AA=FF(a<=LL + k, LL + k, AA))) $ k=1..500:
```

```
>> IF := subs(IF, LL=0, FF=_if):
```

Here, the preceding **IF** program is exactly that generated by calling on 2 instead of 500. Now comparing system times yields:

```
>> time(IF(501));
     550
```

```
>> time(CASE(501));
     283
```

Thus the **case**-statement is twice as fast as the corresponding **if**-statement.

Chapter 3

Programs

In this chapter we present a collection of sample programs written in the MuPAD language.

We start by showing ways and means of increasing the performance and efficiency of MuPAD procedures. The suggested improvements sometimes result in enormous savings with respect to run time. The necessary constructions, however, are not always easy to understand because they are based on knowledge of the internal data structure and the mechanisms of evaluation for MuPAD data. This first section, therefore, is addressed to the experienced user, who works with problems of such complexity that the system is challenged to its limit. Such a user is certainly willing to improve the efficiency of his programs, even if this causes a loss in transparency.

For the beginner of MuPAD programming however, we recommend that whenever there is a choice between transparency of programs and run time savings, he should choose transparency. However, tricks like those presented in the following section will be used for the implementation of the MuPAD libraries.

In the subsequent sections sample programs of a more transparent nature are given. These programs deal with a variety of problems such as writing a user defined differential procedure, or implementing a matching algorithm, or using the domain concept for an extension of data structures.

3.1 Some Tips and Tricks

Whenever, in the following examples, there is a statement about the run time of a procedure, then these are the run times measured by the time command on a Macintosh IIfx. Therefore the results of these measurements are not absolute quantities since they depend on the hardware platform.

We start with a very simple example: the recursively defined procedure for computing the n-th Fibonacci number:

```
>> fib := proc(n)
        begin
          if n<2 then n else fib(n-1)+fib(n-2) end_if;
        end_proc:
```

With respect to simplicity and transparency this program is unsurpassed and, therefore, it is often seen as an example in the manuals of other computer algebra systems. The disadvantage of this program however, is, that even for a two digit number **n fib** calls itself so often that a high run time is the result. In order to see how often fib calls itself we insert a counter. Now we see that for **n = 10** there are 177 calls, and for **n = 18** there are 8361 calls, due to the recursive nature of the procedure. Obviously the number of calls grows exponentially with the size of the argument of **fib** and this leads to an enormous slow down of computation. The computation of **fib(18)** leads to an unbearable run time of 8 seconds.

Here the option **remember** leads to considerable speeding up of computation:

```
>> fib_remember := proc(n)
                     option remember;
                   begin
                     if n<2 then n else fib_remember(n-1)
                              +fib_remember(n-2) end_if;
                   end_proc:
```

This option saves all computed values in the remember table of **fib_remember**. Thus no computation is done twice. This reduces considerably the number of calls of **fib_remember**. Now, for the first computation of **fib_remember(18)** only nineteen calls are necessary, which result in 83 milliseconds of run time, already an improvement by the factor 100. If calling on **fib_remember** for a second time, with the same parameter the run time is hardly measurable; the time command returns 0 milliseconds. This is caused by direct access to those values which were saved in the remember table. However, even this procedure is far from being optimal, and some precautionary measures are necessary. For example, when the user calls **fib_remember(1000)** for the first time then this leads to a breakdown in small computers caused by a stack overflow.

A remarkable improvement is achieved by the following procedure which is even shorter:

```
>> fib_fast := proc(n) local i ;
              begin 0; 1;
                for i from 1 to n-1 do eval(%1+%2) end_for;
              end_proc:
```

It needs exactly 40 milliseconds for the first computation of **fib(18)**, and for **fib_fast(1000)** only 3.5 seconds are necessary. This improvement will be explained later in this section. However, when based on a better mathematical background like, for example, the Binet formula, the run time can again be considerably improved for arguments which are big integers.

MuPAD does not supply a function which, in analogy to the operator for expression sequences, generates sequences of statements. We therefore want to write such a function in the MuPAD language:

```
>> reset():
```

```
>> s_seq := proc(ex, range)
             local last;
             option hold;
             begin
               level(hold(_stmtseq)(op(args(),
                  3..nops(args())), ex$range));
             end_proc:
```

Here the declaration of `last` as a local variable prevents the evaluation of a `last` given within an argument. Thus, any such `last` is considered a command which can be evaluated only outside the environment of the procedure `s_seq`. To ensure this, a `hold` attribute is given to the function `s_seq`. The command in the fifth and sixth line serves basically to convert an expression sequence into a statement sequence; the complicated call is necessary in order to control the evaluation of that expression. The use of further arguments is connected with the behavior described below.

The functionality of this procedure now corresponds to the underline equivalent of the sequence operator. Only, instead of an expression sequence, a statement sequence is generated:

```
>> s_seq(i^2, i=1..5);
     (1; 4; 9; 16; 25)
```

```
>> eval(%);
     25
```

In `s_seq` further arguments may also be used. These then appear as starting elements of the statement sequence generated by this procedure:

```
>> s_seq(i*5, i=1..5, 124, 125);
     (124; 125; 5; 10; 15; 20; 25)
```

Expressions, which are given as a first argument, may contain calls of the function `last`:

```
>> s_seq(%*5, i=1..4, 124, 125);
     (124; 125; last(1)*5; last(1)*5; last(1)*5; last(1)*5)
```

```
>> eval(%);
      78125
```

However, if the call of **s_seq** is followed immediately by an **eval(%)**, then the user has to make sure that, whenever the maximal argument of these **last** calls is **n**, then **n** additional arguments have to be passed to the function **s_seq**. If this precaution is observed then the user has already generated an excellent tool for recursive programming. In case the user wants to avoid the necessity of performing a subsequent evaluation when calling on **s_seq** he can make use of the system function **context**:

```
>> stmt_seq:= proc(ex, ber)
             local   last, A;
             option hold;
             begin
                 A:=level(hold(_stmtseq)(op(args(),
                 3..nops(args()))), eval(ex)$ber), 2);
             context(A);
             end_proc:
```

This function **stmt_seq** has a similar functionality as **s_seq**, however the evaluation of the variable **A** is performed outside the context of the procedure **stmt_seq**. To do this, the system function **context** is available, it first evaluates within the procedure from which it is called, then the result is evaluated in the respective calling environment. The function **stmt_seq** generates an evaluated statement sequence, thus only the evaluation of the final element of that sequence appears as output. This is seen by running through the same examples as those considered for **s_seq**. For example:

```
>> stmt_seq(i*5, i=1..5, 124, 125);
      25
```

Another difference between **stmt_seq** and **s_seq** is that when using **last** in its arguments the elements to which this last refers can be given either before or within the call of **stmt_seq**

```
>> stmt_seq(%*5, i=1..4, 124, 125);
      78125
```

```
>> 124: 125: stmt_seq(%*5, i=1..4);
      78125
```

For example, now the (i+1)-th Fibonacci number is obtained by:

```
>> stmt_seq(%1+%2, i=1..5, 0, 1);
          8.
```

or

```
>> 0: 1: stmt_seq(%1+%2, i=1..150);
              161305314249045814157979073863 49
```

Compared with our fastest procedure this results in a moderate slow down of only 15%, but this seems insignificant compared to the improved transparency of this procedure.

Often the speeding up of procedures, like that of **fib_fast** are based on the efficiency of the sequence operator and the use of the function **last**. This is due to the output of **last** not being evaluated further. A general rule is that evaluation, simplification and expansion lead to a considerable slowing down of programs.

Avoiding these operations, whenever they are unnecessary, results in considerable run time saving. To illustrate this point we compare two procedures which determine how many elements of two lists coincide. In the first case, the list elements are accessed via indexing:

```
>> compare := proc(A, B)
              local m, n, i, k, z;
              begin
                z:=0;
                m:=nops(A);
                n:=nops(B);
                for i from 1 to m do
                  for k from 1 to n do
                    if A[i]=B[k] then
                      z:=z+1
                    end_if
                  end_for
                end_for;
                z;
              end_proc:

>> compare([a, b], [a, g]);
      1
```

In the second procedure the elements of the list are accessed directly by running over the operands of the lists, without using indices. Furthermore, unnecessary evaluations are suppressed by the **val** function:

```
>> compare2 := proc(A, B)
              local a, b, z;
              begin
                z := 0;
                for a in val(A) do
                  for b in val(B) do
                    if a = b then
```

```
                    z := z + 1
                 end_if
              end_for
           end_for;
              z;
         end_proc:
```

```
>> compare2([a, b], [a, g]);
      1
```

Now, comparing two lists, each having 100 elements, the user finds that the second procedure is indeed almost two times faster than the first:

```
>> M := [a.i $ i=1..100]: N := [b.k $ k=1..100]:
```

```
>> time(compare(M, N)); time(compare2(M, N));
      25366
      14050
```

The reason for this is that elements of lists, as well as elements of expression sequences, are evaluated in each access of the list. Of course, the number of performed iterations in both procedures is identical. But in **compare** many more evaluations have to be done than in **compare2**. In **compare**, both lists A and B are accessed $n \cdot m$ times (n and m are the number of elements in A and B respectively), whereas in **compare2**, A is accessed only once, namely in the **for in** loop, and B is accessed n times, where again, n is the number of elements in A.

In interactive mode, wherever possible, the function **last** should be used instead of an expression or identifier. This does not increase the transparency of expressions, however, the advantages with respect to speed is considerable. To illustrate this point we give two statements computing the sum $\sum_{i=1}^{200} a \cdot x^i + b \cdot x^i$ for unassigned a, b and x. First, we proceed by recursive assignment:

```
>> S:=0:
time((for i from 1 to 200 do S:=S+a*x^i + b*x^i  end_for));
      52416
```

The evaluation of this expression is five times slower when compared with an expression where on the right hand side of the assignment the identifier S is replaced by **last**:

```
>> 0:
time((for i from 1 to 200 do % + a*x^i + b*x^i end_for));
      9916
```

However, this observation does not necessarily apply to procedures, as seen by comparing the following examples:

```
>> sum1 := proc(n)
            local i, S;
          begin
            S:=0;
            for i from 1 to n do
              S:=S+a*x^i + b*x^i;
            end_for;
          end_proc:

>> sum2 := proc(n)
            local i, S;
          begin
            0;
            for i from 1 to n do
              % + a*x^i + b*x^i;
            end_for;
          end_proc:
```

Both cases are almost equally fast:

```
>> time(sum1(200));
time(sum2(200));
      9483
      10300
```

However, even for these programs a further improvement, by the factor 10, is possible by use of the sequence operator together with the corresponding underline function:

```
>> sum := proc(n)
            local i;
          begin
              a*x^i + b*x^i $ i = 1..n:
              _plus(%);
          end_proc:
time(sum(200));
      950
```

The combination of sequence operator and underline functions does not always lead to such dramatic savings in run time. To show this we shall compare two programs for the computation of binomials. Programming of the recursion formula with help of a **for**-statement leads to:

```
>> binom1 := proc(n, k)
            local bk;
            begin
              bk:=0;
```

```
          if k>n then bk;
          else  bk:=1;
            case k
            of 0 do bk:=1; break
            of 1 do bk:=n; break
            otherwise
              for i from 1 to k-1 do bk:=bk*(n-i)/i; end_for;
              bk:=bk*n/k;
            end_case;
          end_if;
          bk;
        end_proc:
```

The following program, where the **for**-statement has been replaced by a sequence
operator, is shorter and has the additional advantage that it also works for non
integer values **n**:

```
>> binom2 := proc(n, k)
            local i;
            begin
              if k = 0 then return(1) end_if;
              if k > n then return(0) end_if;
              if k>n div 2 then binom2(n, n-k) end_if;
              _mult(n+1-i$i=1..k);
              %/fact(k);
            end_proc:
```

However, for many arguments this only leads to minimal run time savings compared to the first program, although due to the different structure of the algorithm
there are also arguments for which the savings are considerable. The reason for
this rather modest improvement in average is that most data were integers, which
need only few evaluations and simplifications.

We want to point out the usefulness of the function **val**. As an illustration we
shall consider a procedure for computing the Taylor polynomials. The following
program is already rather polished and has, for a library function, a reasonable
run time:

```
>> taylor := proc(f, eqn, n)
            local a, k, x;
          begin
            x := op(eqn, 1);
            a := op(eqn, 2);
            _plus(eval(subs(f, eqn)),
            eval(subs(diff(f, x $ k ), eqn)) / fact(k)
                    * (x-a)^k $ k = 1..n);
          end_proc:
```

```
>> taylor(sin(x^2), x=0, 20);
     x^2+x^6*(-1/6)+x^10*1/120+x^14*(-1/5040)+x^18*1/362880
```

Most of the tricks mentioned so far have been used: few assignments and the use of the sequence operator and underline functions. However, it is still possible to shorten the run time of this program by a considerable factor (4 for the example above):

```
>> taylor_fast := proc(f, eqn, n)
                local x, y, a, i, xi;
                begin
                  x := op(eqn, 1);
                  y := op(eqn, 2);
                  a:=f; xi:=NIL;
                  (a:=diff(val(a), x)/i*xi) $ i=1..n;
                  _plus(eval(subs((f,level(%,1)), [x=y, xi=x-y])));
                end_proc:
```

```
>> taylor_fast(sin(x^2), x=0, 20);
     x^2+x^6*(-1/6)+x^10*1/120+x^14*(-1/5040)+x^18*1/362880
```

The reason for this considerable improvement is the use of the function **val**, by which unnecessary evaluations are prevented. The run time of this MuPAD function comes close to that of the corresponding system functions of some computer algebra systems.

At the end of the section dealing with run time problems, the evaluation memory of MuPAD should be mentioned. Evaluation memory means that intermediate results are saved until it is logically required that they must be forgotten, in general, until the next assignment takes place. To see the effect of evaluation memory, we shall perform a simple but nevertheless costly computation twice:

```
>> time(fact(10000));
     33183
```

```
>> time(fact(10000));
     0
```

MuPAD is not really as fast as the second computation suggests. The system only remembered the result of the first computation. The evaluation memory provides another good reason for not using too many assignments. By unnecessarily assignments, the data, to which the system has access with more speed, are erased.

3.1.1   A non commutative Product

To demonstrate that functional environments are not worthless toys we want to present a convincing example. Unfortunately, good examples in that respect do necessarily lack transparency.

We want to define a symbol, say **&t**, representing the non commutative product in an arbitrary algebraic structure. For such products we want simplification according to the rules of associativity and distributivity to be automatically performed. In addition, the product sign should have higher binding priority than the usual sign + for addition and the output on screen should be in the form of **A &t B**.

Before realizing the task we would like to mention that MuPAD offers a special notation for binary operators. If for example a function **F** is defined, then the user can write **a &F (g, k)** which MuPAD transforms automatically into **F(a, g, k)**:

```
>> a &F (g, k);
    F(a, g, k)
```

However this does not yet solve our problem since MuPAD only understands this &-notation but does not return its results in this form. Nevertheless the problem is solvable as the following examples demonstrate:

```
>> a &t b;
    a &t b

>> a &t (b+c);
    a &t b + a &t c

>> (a+b) &t (a+c);
    a &t a + b &t a + a &t c + b &t c

>> (a+b) &t (c+d) + e;
    e + a &t c + a &t d + b &t c + b &t d
```

Here, how it was done: associativity was taken from the operator **_exprseq**, where associativity was called flattening. In the functional environment of that operator we change the entries for the output priority, the operator symbol and the name. The result of these replacements we save as a function called **_write**:

```
>> _write := subsop(_exprseq, [1,3] = "_write", [2,2] = 15,
[2,3] = " &t ",[2,4] = "_write"):
```

Now in order to obtain the desired output not only from the input of **_write**, but also from input of the form **a &t b**, we assign to the identifier **t** a suitable procedure:

```
>> t:= proc()
local a,b, i;
begin
    if args(0) > 2 then
        t(args(1), t(op(args(),2..args(0))))
    else
        case args(0)
```

```
            of 1 do args(1); break
            of 2 do a:=args(1); b:= args(2);
                if type(a) = "_plus" then
                    _plus(t(op(a,i),b) $ i=1..nops(a))
                elif type(b)= "_plus" then
                    _plus(t(a,op(b,i)) $ i=1..nops(b))
                else _write(a,b)
                end_if;
                break;
            end_case;
        end_if;
end_proc:
```

This procedure applies the law of distributivity and thereafter passes the result to
the procedure **_write**.

3.1.2   Evalassign and FUNC

In this section we introduce the programs which were used either in the sample
session or in the section 3.1 on Tips and Tricks. Some of these programs are
a bit complex; these are addressed to the experienced user. If, at some point,
the functionality of some program steps are not understood then we recommend
performing experimental runs or analyzing these programs with the debugger or
an increased **PRINTLEVEL** variable.

We already used the operation **FUNC** several times during this tutorial although it
is not a function of the MuPAD kernel. We describe this useful function next. We
start by writing a function, which, if an argument of the form **f(x)**, is given, then
the value 1 is assigned to **f**. And this should happen independent of whether a
value was already assigned to the identifier **f** or not. This is far from being a simple
problem, since assignments of the form **op(argument, 0):=1** are not understood
as assignments to the zero operand, but as assignments to the remember table of
op. More general, our task is to write a function which assigns to the evaluation
of an argument **left**, with evaluation depth **depth**, an arbitrary value **right**:

```
>> evalassign_1 := proc(left, right, depth)
                local EVAL_STMT, stmt, s1, s2, newdepth;
                option hold;                                #a#
            begin
              if  args(0)<>3 then
                error("needs three arguments")
              end_if;
              newdepth:=level(depth);                       #b#
              EVAL_STMT:=FALSE;                              #c#
              stmt:=(s1:=s2);
```

```
                    if domtype(args(2)) = DOM_NIL then
                      stmt:=(subsop(stmt,
                      1=level(left, newdepth+1), 2=NIL))      #d#
                    else
                      stmt:=(subsop(stmt,
                       1=level(left, newdepth+1),
                       2=level(right)))                       #e#
                    end_if;
                    eval(%);
                  end_proc:
```

The arguments are frozen in line **#a#** in order to prevent their premature evaluation. In line **#b#** the `hold` attribute is removed from the last argument. In **#c#** the evaluation of statements inside of expressions is prevented. This is done in order to make this assignment available for substitution. So, in a sense, it is a dummy assignment. To prevent an effect in outer environments, the environment variable `EVAL_STMT` was declared as a local variable. This guarantees that after quitting the procedure its original value is restored. Lines **#d#** and **#e#** serve to differentiate between assigning either `NIL` or another value. This difference reflects the substitution depth of these assignments. Indeed, now, `evalassign_1` has the desired functionality:

```
>> f := NIL: evalassign_1(op(f(x), 0), 15, 1): f;
15
```

```
>> evalassign_1(f, NIL, 0): f;
f
```

A shorter and more elegant solution would have been the following:

```
>> evalassign := proc(left, right, depth)
                 option hold;
                 local dummy;
               begin
               depth:= context(depth);
               if depth = 0
                 then
                 context(subsop(hold(_assign(dummy, dummy)),
                     1=left, 2=right))
               else
                 context(subsop(hold(_assign(dummy, dummy)),
                 1=context(subsop(hold(level(dummy, dummy)),
                 1=left, 2=depth)), 2=right))
               end_if;
             end_proc:
```

This solution of the given problem avoids changing the environment variable
EVAL_STMT.

Now we are in the position to design the auxiliary procedure FUNC which was
already used several times:

```
>> FUNC := proc(argu)
           local newargu, fct, dummy1, dummy2;
         option hold;
         begin
           if type(args(1)) = "_equal" then              #a#
              _exprseq(op(args(1)));
              newargu:=(op(args(1)), NIL, args(i)$i=2..args(0))
           else
              newargu:=args()                             #b#
           end_if;

           fct:=proc(dummy1)                              #c#
           begin
              dummy2;
           end_proc;

           fct:=subsop(fct, 1=op(op(newargu, 1)),
                    4=op(newargu, 2));
           fct:=level(fct);
           domtype(op(newargu, 1));
           if % = DOM_EXPR or % = DOM_ARRAY then          #d#
              op(newargu, [1, 0]);
              if domtype(level(last(1))) = DOM_FUNC then   #e#
                 return("Please, do not redefine
                     functional environments");
              else
                 fct := subsop(fct, 6=op(newargu, [1,0]));
                 evalassign(op(newargu, [1,0]), fct, 1);    #f#
              end_if;
           else
              fct;
           end_if;
         end_proc:
```

Before describing the program steps we should study its functionality. The procedure FUNC accepts an equation as argument as well as an expression sequence:

```
>> FUNC(f(x)=x^7): f;
proc(x) begin x^7 end_proc
```

```
>> FUNC(g(x), x^9): g;
proc(x) begin x^9 end_proc
```

If an expression sequence is given as an argument and the first expression is not a function call, then only the function corresponding to the second argument is returned:

```
>> FUNC(x, x^9);
proc(x) begin x^9 end_proc
```

In line #a# it is determined if the argument is an equation or not. In case of an affirmative answer the input is changed into an expression sequence which is assigned to the parameter **newargu**. If the input was already in the form of an expression sequence, then in line #b# the old arguments are assigned to that parameter. In line #c# a dummy procedure is created in which the operands resulting from the decomposition of the function call are inserted one after another. To do this we need to prevent evaluation by setting a **hold** attribute. In line #f# the auxiliary procedure **evalassign** is used in order to assign the function name which, in general, was extracted as zero operand of the left side of the argument to the new function. Additionally a safeguard against overwriting the system function was introduced in this procedure.

When writing the function **get_ufunc**, which was used in order to inquire about the set of underline functions, we used a concatenation of strings:

```
>> get_ufunc := proc() local i, ufunc;
               begin
                 ufunc := [];
                 for i in anames(0) do
                    if strmatch("".i, "_\*") then
                      ufunc := append(ufunc, i)
                    end_if
                 end_for:
                 ufunc;
               end_proc:
```

Firstly, by use of **anames**, the set of internal MuPAD functions was determined. For this set, evaluation had to be prevented, because this leads in older MuPAD versions to a set of confusing functional environments. In order to do this, we firstly converted the names by concatenation into strings. From these strings we singled out those which start with an underline.

3.2   More sample Programs

In order to demonstrate the flexibility and the broad spectre of MuPAD's applications in this section we present further programs from different areas of applications.

3.2.1   A user defined differential Procedure

A program for differentiation is easily written:

```
>> der := proc()
           local d, n;
           option remember;
        begin
         d:=type(args(1));
         case d
             of DOM_INT do
             of DOM_RAT do
             of DOM_FLOAT do
             of DOM_COMPLEX do
             of DOM_IDENT do
               if args(1)=args(2) then 1 else 0 end_if;
               break
             of "_mult" do
               _plus(der(op(args(1), n), args(2))*
                 subsop(args(1), n=1) $ n=1..nops(args(1)));
               break
             of "_plus" do
               _plus(der(op(args(1), n),
                   args(2)) $ n=1..nops(args(1)));
                 break
             of "_power" do op(args(1), 1);
               op(args(1), 2);
               args(1)*ln(last(2))*der(last(1),
                   args(2))+last(1)*der(last(2),
                   args(2))*last(2)^(last(1)-1);
               break
             otherwise
               _par(text2expr(d))(op(args(1)))*
                           der(op(args(1), 1), args(2))
          end_case;
          eval(last(1));
        end_proc:

>> der(x^5, x);
     x^4*5

>> der(x+x*64, x);
     65
```

When calling this program, the first argument should be the expression of which
the derivative is to be computed, and the second argument is the variable to

which differentiation is performed. The first case, which only comes into effect when the first argument is an identifier, asks whether or not this identifier is equal to the variable of differentiation. If this is the case then the result is **1**. Subsequently, the product rule, the sum rule and the rule for differentiation of powers are implemented in the second to the fourth **case** statement.

Unfortunately, this program for differentiation is not yet able to calculate derivatives of well known functions:

```
>> der(sin(x), x);
     _par(sin)(x)
```

In order to circumvent this disability it is only necessary to teach the function **_par** the derivative of the sine function:

```
>> _par(sin) := cos:
```

Then all derivatives of expressions containing the sine function are more than easy:

```
>> der(sin(x)*sin(x^(1/2)), x);
     cos(x)*sin(x^(1/2))+x^(-1/2)*sin(x)*cos(x^(1/2))*1/2
```

Of course, teaching the derivative of only the sine function is not yet sufficient, the mathematical expert knowledge about derivatives has to be extended to all other functions implemented in the MuPAD kernel:

```
>> _par(sinh):=cosh:

>> _par(cos):=FUNC([x], -sin(x)):

>> _par(cosh):=sinh:

>> _par(tan):=FUNC([x], cos(x)^(-2) ):

>> _par(tanh):=FUNC([x], cosh(x)^(-2)):

>> _par(asin):=FUNC([x], (1-x^2)^(-1/2)):

>> _par(asinh):=FUNC([x], (1+x^2)^(-1/2)):

>> _par(acos):=FUNC([x], -(1-x^2)^(-1/2)):

>> _par(acosh):=FUNC([x], (x^2-1)^(-1/2)):

>> _par(atan):=FUNC([x], (1+x^2)^(-1)):

>> _par(atanh):=FUNC([x], (1-x^2)^(-1)):

>> _par(exp):=exp:
```

```
>> _par(sqrt):=FUNC([x], (2*(x))^(-1/2)):

>> _par(ln):=FUNC([x], (x)^(-1)):
```

And now we have a program which really performs all desired derivatives:

```
>> der(x*sin(atanh(x+x^5))+x^2, x);
   x*2+sin(atanh(x+x^5))+x*cos(atanh(x+x^5))*\
   (x^4*5+1)/(-(x+x^5)^2+1)

>> der(%, x);
   x^4*cos(atanh(x+x^5))/(-(x+x^5)^2+1)*20+\
   cos(atanh(x+x^5))*(x^4*5+1)/(-(x+x^5)^2+1)*2-\
   x*sin(atanh(x+x^5))*(x^4*5+1)^2*(-(x+x^5)^2+1)^\
   (-2)+x*(x+x^5)*cos(atanh(x+x^5))*(x^4*5+1)^2*\
   (-(x+x^5)^2+1)^(-2)*2+2
```

3.2.2 Polynomials

As already mentioned in section 2.5 polynomials are represented by a special data structure. More than 20 functions are available for working with them. Some of them are used in the following examples.

3.2.2.1 Heuristic GCD

For computing the greatest common divisor of two polynomials over the integers, the heuristic method listed below is quite fast. Both polynomials are assumed to be primitive. The result is either the greatest common divisor of the polynomials or **FAIL**. In the univariate case, the algorithm is quite fast and reliable. In the multivariate case, the method should not be used for polynomials with more than five indeterminates.

```
>> heu_gcd:= proc(a, b)
local x, xv, g, tries, bound;
begin
   x:= op(a,[2,1]);
   bound := max(degree(a,x), degree(b,x));
   xv := 2*min(norm(a), norm(b)) + 2;

   for tries from 1 to 6 do
      if strlen("".xv) * bound > 5000 then return(FALSE) end_if;
      if nops(op(a,2)) = 1 then
            g := igcd(evalp(a,x=xv), evalp(b,x=xv))
      else
            g := heu_gcd(evalp(a,x=xv), evalp(b,x=xv));
```

```
                if g = FALSE then return(FALSE) end_if
        end_if;
        if g <> FAIL then
                g := genpoly(g, xv, x);
                if divide(a, g, hold(Exact)) <> FAIL then
                    if divide(b, g, hold(Exact)) <> FAIL then
                        return(g)
                    end_if
                end_if
        end_if;
        xv := trunc(xv * 73794 / 27011);
    end_for;
    FAIL;
end_proc:

>> p1 := poly(x^3 - 6*x^2 + 11*x - 6, [x]):
p2 := poly(x^3 - 11*x^2 + 38*x - 40, [x]):

>> heu_gcd(p1, p2);
    poly(x - 2, [x])

>> p1 := poly(x^2*y*z*6 + x*y^2*z^2*3 + x*6 + y*z*3, [x, y, z]):
p2 := poly(x*y^2*z^2*3 + x*y^2*z*3 + y*z*3 + y*3, [x, y, z]):

>> heu_gcd(p1, p2);
    poly(x*y*z*3 + 3, [x, y, z])
```

3.2.2.2 Orthogonal Polynomials

The Chebyshev polynomials $T(n, x)$ of the first kind are defined by the recurrence
relation:

$$
\begin{aligned}
T(0, x) &= 1 \\
T(1, x) &= x \\
T(n, x) &= 2xT(n - 1, x) - T(n - 2, x), \text{ für } n > 1
\end{aligned}
$$

From this, it is not very difficult to write a MuPAD program to compute them:

```
>> Chebyshev := proc(n,x)
local T;
begin
    case n
      of 0 do
        T := poly(1,[x]);
        break;
```

```
      of 1 do
         T := poly(x,[x]);
         break;
      otherwise
         T := poly(2 * x,[x]) * Chebyshev(n-1,x) - Chebyshev(n-2,x);
   end_case;
   T;
end_proc:

>> Chebyshev(4,x);
      poly(x^4*8+x^2*(-8)+1, [x])

>> Chebyshev(8,x);
      poly(x^8*128+x^6*(-256)+x^4*160+x^2*(-32)+1, [x])
```

With the knowledge we gained in section 3.1, this program can be improved substantially. The number of recursive procedure calls can be reduced drastically by setting the option **remember**. However, with respect to run time, procedure calls are rather expensive. So a further improvement is obtained by the following iterative version of the procedure:

```
>> Cheby_iter := proc(n, x)
      local i;
      begin
        poly(1,[x]); poly( x, [x]);
        for i from 1 to n-1 do
          poly(2*x, [x]) * eval(%1) - eval(%2)
        end_for;
      end_proc:

>> time(Chebyshev(8,x)); time(Cheby_iter(8,x));
      1316
      216
```

3.2.3   Combinatorics: Permutations

A frequent problem associated with a set is the determination of all permutations. A simple recursive procedure for this problem follows. The list has been chosen as the elementary data structure. The solution, i.e. the set of all permutations is also returned in form of a list. The function **permute** works with a local recursive procedure **perm**. This runs through the entire list in a loop. In each run one element of the list is deleted and with this decreased list **perm** calls itself recursively. The abort critieria of the recursion is the **if**-enquiry at the beginning. If the list contains only two elements the solution is directly determined, without any further recursive calls, and returned.

```
>> permute:= proc(list)
local perm;
begin
   perm := proc(l)
   local i, elem, result, res, rl, lcopy;
   begin
      if nops(l) = 2
         then return([[op(l,1), op(l,2)], [op(l,2), op(l,1)]])
      end_if;
      result := []; lcopy := l;
      for i from 1 to nops(l) do
         elem := l[i]; l[i] := NIL;
         res := perm(l);
         for rl in res do
            result := append(result, [elem].rl)
         end_for;
         l := lcopy
      end_for;
      result;
   end_proc;

   perm(list);
end_proc:

>> permute([1,2,3,4]);
   [ [1,2,3,4], [1,2,4,3], [1,3,2,4], [1,3,4,2], [1,4,2,3],
     [1,4,3,2], [2,1,3,4], [2,1,4,3], [2,3,1,4], [2,3,4,1],
     [2,4,1,3], [2,4,3,1], [3,1,2,4], [3,1,4,2], [3,2,1,4],
     [3,2,4,1], [3,4,1,2], [3,4,2,1], [4,1,2,3], [4,1,3,2],
     [4,2,1,3], [4,2,3,1], [4,3,1 2], [4,3,2,1] ]
```

The objects to be permutated are not only limited to numbers, but can be of any
type:

```
>> permute([Apple, "Peach", Pear]);
   [ [Apple,"Peach",Pear], [Apple,Pear,"Peach"],
     ["Peach",Apple,Pear], ["Peach",Pear,Apple],
     [Pear,Apple,"Peach"], [Pear,"Peach",Apple] ]
```

3.2.4   Sorting

One of the most frequently operations carried out on data is sorting. A simple
method for sorting a list of integers is *Binary insertion*, an improved version of
the well-known *Straight insertion* (sorting by direct insertion). A corresponding
MuPAD program could resemble this one:

```
>> bininsert := proc(list)
local i, j, l, r, m, x;
begin
   for i from 2 to nops(list) do
      x := list[i]; l := 1; r := i -1;
      while l <= r do
         m := (l + r) div 2;
         if x < list[m]
            then r := m-1
            else l := m + 1
         end_if;
      end_while;
      for j from i-1 downto l do
         list[j+1] := list[j]
      end_for;
      list[l] := x;
   end_for;
   list;
end_proc:

>> bininsert([10,8,9,6,7,4,5,2,3,1]);
   [1, 2, 3, 4, 5, 6, 7, 8, 9, 10]
```

3.2.5   Graphs

MuPAD is also perfectly suited for the implementation of algorithms on graphs.
Due to the multitude of data structures and the powerful operations which can
be used with them, algorithms for graphs can be implementiert very easily. These
algorithms often contain instructions of the form „for each edge do ...“ or
„for each node do ...“. As examples we would like to introduce two graph
algorithms. The first determines the length of the shortest paths in a directed
graph with a distance function. The second algorithm determines a matching of
maximal cardinality in a bipartite graph.

3.2.5.1   Shortest Paths

We are given a directed graph $G = (V, E)$ with the distance function $c : E \to \mathbb{R}$,
i.e. $c(e)$ is the length of the edge e. We presuppose that G contains no circles with
negative length. We search for a table d, so that $d[i, j]$ is the length of the shortest
path between node i and node j. The routine **short_path** presented below fulfils
this task. As its parameters it expects the node set V and the distance function c
in form of a table.

```
>> short_path := proc(V, c)
local i, j, k, d;
begin
   for i in V do
      for j in V do
         if i <> j
            then d[i,j] := c[i,j]
            else d[i,j] := 0
         end_if;
      end_for;
   end_for;
   for k in V do
      for i in V do
         for j in V do
            d[i,j] := min(d[i,j], d[i,k] + d[k,j])
         end_for;
      end_for;
   end_for;
   d;
end_proc:
```

But, which concept is behind this algorithm? We would like to examine it in the following. In G we regard the shortest path between the node i and the node j. If k is a node on this path, so must the paths from i to k and from k to j also be shortest paths. If k is the node on this path with the highest index then the path from i to k is one of the shortest paths. On this path no node has an index higher than $k - 1$. The same is true for the path from k to j. To determine a shortest path from i to j the node k with the highest index must be determined. Then the shortest paths from i to k and k to j must be determined, whereby neither of these paths can contain a node with an index higher than k. If $d^k(i, j)$ is the length of the shortest path from i to j, which contains no node with an index higher than k, then:

$$d(i, j) \ = \ \min(\min_{1 \leq k \leq n} (d^{k-1}(i, k) + d^{k-1}(k, j)), c(i, j))$$

A shortest path from i to j, which passes through no nodes with a higher index as k, can either pass through k or not. Thus in the first case $d^k(i, j) = d^{k-1}(i, k) + d^{k-1}(k, j)$ is valid and the second case, in which no node has an index higher than $k - 1$, results in $d^k(i, j) = d^{k-1}(i, j)$. This leads to the following recurrence equation for $d^k(i, j)$:

$$d^k(i, j) \ = \ \min(d^{k-1}(i, j), d^{k-1}(i, k) + d^{k-1}(k, j)), \quad k \geq 1$$

As an example for this we shall examine the following graph:

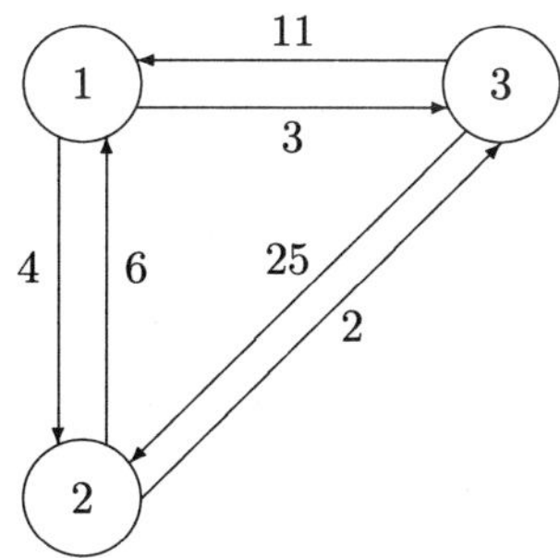

The nodes are stored in a set, the weighted edges in a table. An explicit storing of the edges is not necessary.

```
>> V := {1, 2, 3}:
c[1,1] := 0: c[1,2] :=  4: c[1,3] := 11:
c[2,1] := 6: c[2,2] :=  0: c[2,3] :=  2:
c[3,1] := 3: c[3,2] := 25: c[3,3] :=  0:

>> short_path(V,c);
     table(
         (1, 1) = 0, (1, 2) = 4, (1, 3) = 6,
         (2, 1) = 5, (2, 2) = 0, (2, 3) = 2,
         (3, 1) = 3, (3, 2) = 7, (3, 3) = 0
     )
```

A further example:

```
>> V := {1, 2, 3, 4, 5}:
c := table(
   (1,1) =  0, (1,2) =  2, (1,3) =  4, (1,4) = 25, (1,5) = 3,
   (2,1) =  2, (2,2) =  0, (2,3) =  8, (2,4) = 25, (2,5) = 1,
   (3,1) =  6, (3,2) =  2, (3,3) =  0, (3,4) =  4, (3,5) = 3,
   (4,1) =  1, (4,2) = 25, (4,3) = 25, (4,4) =  0, (4,5) = 5,
   (5,1) = 25, (5,2) = 25, (5,3) = 25, (5,4) =  1, (5,5) = 0
):

>> short_path(V,c);
     table(
         (1, 1) = 0, (1, 2) = 2, (1, 3) = 4, (1, 4) = 4, (1, 5) = 3,
         (2, 1) = 2, (2, 2) = 0, (2, 3) = 6, (2, 4) = 2, (2, 5) = 1,
         (3, 1) = 4, (3, 2) = 2, (3, 3) = 0, (3, 4) = 4, (3, 5) = 3,
         (4, 1) = 1, (4, 2) = 3, (4, 3) = 5, (4, 4) = 0, (4, 5) = 4,
         (5, 1) = 2, (5, 2) = 4, (5, 3) = 6, (5, 4) = 1, (5, 5) = 0
     )
```

3.2.5.2 Matching

Let $G = (V, E)$ be an undirected graph. A *matching* is a set of edges where each
node is incident to no more than one edge. The matching problem for a graph
consists of finding a match of maximal cardinality.

Let G be a graph and M a matching in G. The edges of M are called *matched*
edges the others are *free* edges. If $[v, u]$ is a matched edge then is u the "mate"
of v. Nodes that are not incident to a matched edge are also called *free*. Analogue to
the edges all other nodes are also called matched. To find a matching alternating
paths play an important role. A path $p = [u_1, u_2, \ldots, u_k]$ is called *alternating* when
the edges $[u_1, u_2]$, $[u_3, u_4], \ldots$, $[u_{2j-1}, u_{2j}]$ are free and the edges $[u_2, u_3]$, $[u_4, u_5]$,
$\ldots$, $[u_{2j}, u_{2j+1}]$ are matched. An alternating path $p = [u_1, u_2, \ldots, u_k]$ is called an
augmenting path when u_1 and u_k are free nodes.

If P is a set of edges on an augmenting path $p = [u_1, u_2, \ldots, u_{2k}]$, with respect to a
matching M, then $M' = M \oplus P$ is a matching with the cardinality $|M|+1$. Thereby
$S \oplus T = (S - T) \cup (T - S)$ signifies the symmetrical difference of sets. With this
an existing matching can be enlarged in a simple manner. Hereby is a matching
M in a graph G a maximum matching if and only if there is no augmenting path
in G with respect to M.

With this the basic concept of the algorithm is clearly defined: Begin with any
matching (e.g. an empty one) and repeatedly search for augmenting paths until
no more such paths exist. This search for extended paths is especially simple in
bipartite graphs. In the following MuPAD program this theory has been converted
into an executable program:

```
>> bipartite_matching:=proc(V,U,EDGE)
local v, erg, A, Q, mate, exposed, label, init,
      build_Q, build_A, augment, loop;
begin

    init := proc(V, U, EDGE)
    local v;
    begin
       for v in V do exposed[v] := 0; end_for;
       A := {}; build_A(V,U,EDGE); build_Q(V,U,EDGE);
       loop(V,U,EDGE);
    end_proc:

    build_A := proc(V,U,EDGE)
    local v, u, edge;
    begin
       for edge in EDGE do
           v := op(edge,1); u := op(edge,2);
           if mate[u] = 0 then exposed[v] := u
```

```
            elif mate[u] <> v then A := A union {[v,mate[u]]}
         end_if;
      end_for;
   end_proc:

   build_Q := proc(V,U,EDGE)
   local v;
   begin
      Q := {};
      for v in V do
         label[v] := 0;
         if mate[v] = 0 then Q := Q union {v} end_if;
      end_for;
   end_proc;

   augment := proc(v)
   begin
      if label[v] = 0
         then mate[v] := exposed[v]; mate[exposed[v]] := v;
         else exposed[label[v]] := mate[v]; mate[v] := exposed[v];
            mate[exposed[v]] := v; augment(label[v]);
      end_if;
   end_proc;

   loop := proc(V,U,EDGE)
   local v;
   begin
      while Q <> {} do
         v := op(Q,1); Q := Q minus {v};
         if exposed[v] <> 0  then augment(v); init(V,U,EDGE);
            else for vs in V do
               if label[vs] = 0
                  then for edge in A do
                     if edge=[v,vs]
                        then label[vs] := v; Q := Q union {vs}
                     end_if;
                     end_for;
               end_if;
               end_for;
         end_if;
      end_while;
   end_proc;

# begin main program #
```

```
    for v in (V union U) do mate[v]:= 0; end_for;
    init(V, U, EDGE); erg := {};
    for v in V do
        if mate[v] <> 0 then erg := erg union {[v,mate[v]]} end_if;
    end_for;
    erg;
end_proc:
```

Before the algorithm is explained we should look at an example:

```
>> V:={v1,v2,v3,v4,v5,v6}: U:={u1,u2,u3,u4,u5,u6}:
EDGE:={[v1,u1], [v1,u2], [v1,u4], [v2,u2], [v2,u6], [v3,u2],
       [v3,u3], [v4,u3], [v4,u5], [v4,u6], [v5,u3],
       [v5,u4], [v5,u5], [v5,u6], [v6,u2], [v6,u5]}:
```

The example is equivalent to the following graph:

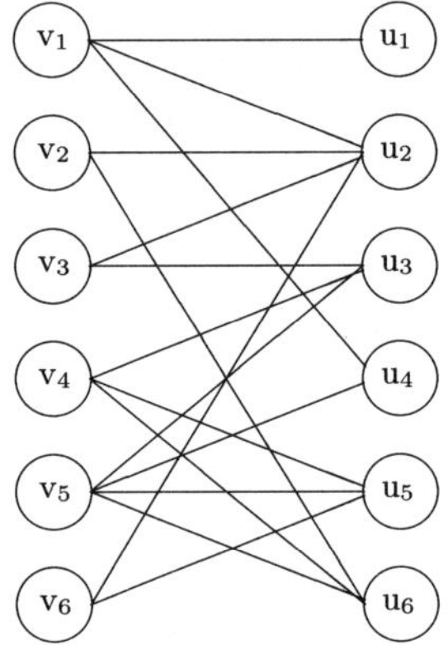

```
>> bipartite_matching(V,U,EDGE);
       { [v1, u1], [v3, u3], [v6, u2], [v2, u6], [v5, u4], [v4, u5] }
```

The algorithm expects the bipartite graph in form of two sets V and U (together these form the set of nodes of the graph) and an edge set E. Edges are represented by two-element lists. A simplified form of breadth-first search is used to search for alternating paths. The table `mate` contains $|V|+|U|$ entries and is used to represent the current matchings. For $v \in V$ it is true that: `exposed[v]` is an element of U, free and adjacent to v. If such a node does not exist then `exposed[v]` = 0. Hereby we have found an extended path for $v \in V$ with `exposed[v]` $\neq 0$. `label`, `A` and `Q` are used for the execution of the actual search. `A` contains the directed auxillary graphs in which a breadth-first search is carried out.

3.2.6 Domains

In this section we would like to give further examples for working with domains. The first example describes a domain generator: the parametrized definition of a new domain. In the second example the real numbers are extended by the symbols ∞ und $-\infty$.

3.2.6.1 Domain Generator

In section 2.8 we exemplarily presented a domain of the ring of remainder classes modulo 7. In this section we want to generalize this example by giving a domain structure that returns the domain $\mathbb{Z}/n\mathbb{Z}$ of the ring of remainder classes modulo n dependant on a positive integer n. Apart from being a further example for working with domains this is also an example for the effective application of the program manipulation possibilities in MuPAD. All necessary routines in the domain, e.g. addition, multiplication etc., are generated during run time, i.e. programs are generated. The procedure for this is clearly defined. The domain generator contains generic forms of these routines. During the run time the parameter specific values are then substituted.

```
>> moduloring := proc(n)
begin
   if args(0) <> 1 then
      error("Wrong number of arguments");
   end_if;
   if domtype(n) <> DOM_INT then
      error("Wrong type of argument");
   end_if;
   if n <= 0 then
      error("Wrong argument");
   end_if;

   mdom := domain():

   #  addition  #
   mdom::_plus :=
      proc(x)
      local i;
      begin
         x := extop(x, 1);
         for i from 2 to args(0) do
            x := (x + extop(args(i), 1)) mod n;
         end_for;
         new (mdom, x);
```

```
      end_proc;

#  negation  #
mdom::negate :=
   proc(x)
   begin
      new (mdom, -extop(x, 1) mod n);
   end_proc;

#  subtraction  #
mdom::substract :=
   proc(x,y)
   begin
      new (mdom, (extop(x, 1)-extop(y, 1)) mod n);
   end_proc;

#  multiplication  #
mdom::_mult :=
   proc(x)
      local i;
   begin
      x := extop(x, 1);
      for i from 2 to args(0) do
         x := (x * extop(args(i), 1)) mod n;
      end_for;
      new (mdom, x);
   end_proc;

#  inverse wrt. multiplication  #
mdom::invert :=
   proc(x)
      local a;
   begin
      a := 1/extop(x, 1);
      new (mdom, a mod n);
   end_proc;

#  binary division  #
mdom::divide :=
   proc(x, y)
   begin
      new (mdom, extop(x, 1)/extop(y, 1) mod n);
   end_proc;
```

```
# exponentiation #
mdom::_power :=
   proc(x,y)
      local h, list;
   begin
      if domtype(y) <> DOM_INT then
         error("Wrong exponent");
      end_if;

      if y=0 then
         return (one);
      end_if;
      if y<0 then
         x := inverse(x);
         y := -y;
      end_if;

      list := [];
      while y<>1 do
         list := append(list, y mod 2);
         y := y div 2;
      end_while;
      y := nops(list);
      h := x;
      while y<>0 do
         h := __mult(h, h);
         if list[y]=1 then
            h := __mult(x, h);
         end_if;
         y := y-1;
      end_while;
      h;
   end_proc;

# definition of a new domain element #
mdom::new :=
   proc(x)
   begin
      if args(0) <> 1 then
         error("Wrong number of arguments");
      end_if;
      if domtype(x) <> DOM_INT then
         error("Wrong type of argument");
      end_if;
```

```
         new (mdom, x mod n);
      end_proc;

   #  output of a domain element  #
   mdom::print :=
      proc(x)
      begin
         extop(x, 1);
      end_proc;

   #  converting functions  #
   mdom::convert :=
      proc(x)
      begin
         if domtype(x) = DOM_INT then
            mdom::new(x);
          else
            FAIL
          end_if;
      end_proc :

mdom::expr :=
   proc(x)
   begin
      extop(x,1);
   end_proc :

   #  neutral elements  #
   mdom::zero := new(mdom, 0);
   mdom::one := new(mdom, 1);

   #  domain name  #
   mdom::name := "Z".n;

   subs(mdom, hold(mdom) = mdom);
   subs(mdom, hold(inverse) = mdom::invert);
   subs(mdom, hold(n) = n);
   subs(mdom, hold(__mult) = mdom::_mult);
   subs(mdom, hold(one) = mdom::one);

   mdom;
end_proc:
```

The five **subs** calls at the end of the domain generator are probably puzzling.

It is suprising that the function calls are not in an assignment. The reason for this is to be seen in a "reference effect", that is attached to the domains. With the modification of already existing domains the changes are not carried out on the copy of the domain, which would produce a new datum, but the changes are executed on the domain itself. Therefore the **subs** calls above do not need to be contained in an assignment. Naturally this reference effect is found in all data which contain a domain as a subdatum. These are especially the elements of a domain.

The **subs** calls are used to accelerate the domain methods. By substituting the identifiers **mdom**, **inverse**, **__mult** and **one** with their values an unnecessary step, which would have normally cost execution time, is omitted. The substitution of the identifier **n** with the current parameter **n** converts the generic domain definition into the desired domain $\mathbb{Z}/n\mathbb{Z}$.

Finally some more examples:

```
>> Z7 := moduloring(7);
      Z7

>> a := Z7::new(5); b := Z7::new(11);
      5
      4

>> a + b; a * b; a - b; a / b;
      2
      6
      1
      3

>> a ^ 6;  b ^ 4;
      1
      4
```

3.2.6.2 Extension of the real numbers by ∞ and $-\infty$

In some applications it is useful to integrate symbols, such as ∞, in the usual numbers. We would like to demonstrate how this can be effected with the help of domains in the following example. Here we extend the real numbers by ∞ and $-\infty$. We achieve this as follows: We generate a new domain **Rinf** which contains the usual numbers as elements. Furthermore there are two singular elements **Rinf::infinity** (for ∞) and **Rinf::minfinity** (for $-\infty$). The domain methods must test if these elements are contained in their arguments. If this is not the case then the usual number operations can be used. Otherwise the operations must be executed with regard to the two special elements.

Firstly a new domain is generated by calling `domain`. This domain gets a name and the methods for defining a new domain element, type testing and output are determined. Then the zero and unit element, along with the two elements for ∞ and $-\infty$, are defined.

```
>> Rinf := domain():
Rinf::name := "Rinf":

Rinf::new := proc(x)
begin
   new(Rinf,x);
end_proc:

Rinf::type := proc(x)
begin
   Rinf;
end_proc:

Rinf::domtype := proc(x)
begin
   Rinf;
end_proc:

Rinf::print := proc(x)
local expr, infinity;
begin
   if x = Rinf::infinity then
      expr := hold(infinity);
   elif x = Rinf::minfinity then
      expr := hold(-infinity);
   else
      expr := subs(hold(Rinf(q)), hold(q)=extop(x, 1));
   end_if;
   expr;
end_proc:

Rinf::zero := Rinf::new(0):
Rinf::one := Rinf::new(1):
Rinf::minfinity := Rinf::new(hold(MInfinity)):
Rinf::infinity := Rinf::new(hold(Infinity)):
```

Now the basic operations $+$, $*$, $\char94$ are overloaded. Therein it is tested, if ∞ and/or $-\infty$ are among the parameters. If this is the case then the corresponding result is returned. Otherwise the usual operation is executed on the numbers.

```
>> Rinf::_plus := proc()
```

```
local i, L, c1, c2;
begin
   L := [args()]; c1:= 0; c2 := 0;
   while (i := contains(L, Rinf::minfinity)) <> 0 do
      L[i] := NIL; c1 := c1 + 1;
   end_while;
   while (i := contains(L, Rinf::infinity)) <> 0 do
      L[i] := NIL; c2 := c2 + 1;
   end_while;

   if c1 > c2 then return(Rinf::minfinity)
   elif c2 > c1 then return(Rinf::infinity)
   elif c1 = c2 and c1 > 0 then return(Rinf::minfinity)
   else
      i := NIL;
      extop(args(i), 1) $ i = 1..args(0);
      Rinf::new(_plus(%));
   end_if;
end_proc:

Rinf::_mult := proc()
local i, L, sign, c1, c2;
begin
   if contains([args()], Rinf::zero) <> 0
      then return (Rinf::zero)
   end_if;
   L := [args()]; sign := 1; c1:= 0;
   while (i := contains(L, Rinf::minfinity)) <> 0 do
      L[i] := NIL; sign := sign * (-1); c1 := c1 + 1;
   end_while;
   c2 := 0;
   while (i := contains(L, Rinf::infinity)) <> 0 do
      L[i] := NIL; c2 := c2 + 1;
   end_while;
   if c1 > 0 or c2 > 0 then
      if sign = -1
         then return(Rinf::minfinity)
         else return(Rinf::infinity)
      end_if;
   end_if;
   i := NIL;
   extop(args(i), 1) $ i = 1..args(0);
   Rinf::new(_mult(%));
end_proc:
```

```
Rinf::_power := proc(x,y)
begin
   if x < Rinf::zero
      then return(FAIL)
   elif x = Rinf::infinity or y = Rinf::infinity
      then return (Rinf::infinity)
   elif y = Rinf::minfinity
      then return(Rinf::one)
   else
      return (Rinf::new(_power(extop(x,1), extop(y,1))));
   end_if;
end_proc:
```

Now the inverse elements with respect to addition and multiplication are defined.

```
>> Rinf::negate := proc(x)
begin
   case x
      of Rinf::infinity do
         return (Rinf::minfinity);
         break;
      of Rinf::minfinity do
         return (Rinf::infinity);
         break;
      otherwise
         return (Rinf::new(-extop(x,1)));
   end_case;
end_proc:
```

```
Rinf::invert := proc(x)
begin
   case x
      of Rinf::infinity do
         return (Rinf::zero);
         break;
      of Rinf::minfinity do
         return (Rinf::zero);
         break;
      otherwise
         return (Rinf::new(1/extop(x,1)));
   end_case;
end_proc:
```

Only the embedding of ∞ and $-\infty$ in the relational operators $<$ and $\leq$ is now missing. In this case, once again a simple case distinction is necessary.

```
>> Rinf::_less := proc(x,y)
begin
   if x = Rinf::minfinity and y = Rinf::minfinity
      then return (FALSE)
   elif x = Rinf::minfinity
      then return (TRUE)
   elif y = Rinf::minfinity
      then return (FALSE)
   elif x = Rinf::infinity
      then return(FALSE)
   elif y = Rinf::infinity
      then return (TRUE)
   else return (bool(_less(extop(x,1), extop(y,1))));
   end_if;
end_proc:

Rinf::_leequal := proc(x,y)
begin
   if x = Rinf::minfinity
      then return (TRUE);
   elif x = Rinf::infinity and y = Rinf::infinity
      then return (TRUE);
   elif x = Rinf::infinity
      then return (FALSE);
   elif y = Rinf::minfinity
      then return(FALSE)
   else return (bool(_leequal(extop(x,1), extop(y,1))));
   end_if;
end_proc:
```

Finally the definition of the system functions min and max.

```
>> Rinf::min := proc()
local i, L;
begin
   if contains([args()], Rinf::minfinity) <> 0
      then return (Rinf::minfinity)
      else
         L := [args()];
         while (i := contains(L, Rinf::infinity)) <> 0 do
            L[i] := NIL;
         end_while;
         i := NIL;
         L := op(L[i], 1) $ i = 1..nops(L);
         Rinf::new(min(L));
```

```
      end_if;
end_proc:

Rinf::max := proc()
local i, L;
begin
   if contains([args()], Rinf::infinity) <> 0
      then return (Rinf::infinity)
      else
         L := [args()];
         while (i := contains(L, Rinf::minfinity)) <> 0 do
            L[i] := NIL;
         end_while;
         i := NIL;
         L := op(L[i], 1) $ i = 1..nops(L);
         Rinf::new(max(L));
   end_if;
end_proc:
```

By overloading the system functions `_plus`, `_mult` etc. the domain methods are
automatically called when domain elements are added, multiplied etc. The element
`Rinf::minfinity` is always printed as - `infinity` so that the existence of this
element remains hidden most of the time.

```
>> r1 := Rinf::new(2); r2 := Rinf::new(4);
infinity := Rinf::infinity;
     Rinf(2)
     Rinf(4)
     infinity

>> type(r1); domtype(-infinity);
     Rinf
     Rinf

>> r1 * infinity; min(infinity, r1, r2);
max(infinity, r1, -infinity, r2);
     infinity
     Rinf(2)
     infinity

>> bool(-infinity < r2);  bool(infinity < r1);
     TRUE
     FALSE
```

By the definition of a domain method `func_call` the domain elements can also be
used as functions. If domain elements occur in a functional environment then a

method `func_call` is searched for in the domain during evaluation. If it is present then this routine is called with the functional environment parameters.

We shall interpret such a function call as the call of a constant function for the domain `Rinf` defined above:

```
>> Rinf::func_call := proc()
begin
      args(1);
end_proc:

>> r1(x,y,z); infinity(2);
      Rinf(2)
      infinity
```

In the definition of the domain `Rinf` the identifier `Rinf` is explicitly used. This has the disadvantage that this definition is only valid when the domain is also stored under the name `Rinf`. If another name is used to store the domain the user can no longer correctly use the domain. In order to correct this defect we can replace the identifier `Rinf` with the domain itself within the domain. This takes place, for instance, by: 

```
>> subs(Rinf, hold(Rinf) = Rinf):
```

The user should note that with this a recursive data structure is produced whose tree, logically seen, is infinitely deep!

3.2.7 Data Manipulation: Flattening

In MuPAD the procedure for *flattening* is the application of the associative law. In many system functions flattening belongs implicitly to evaluation, like for instance in addition where $a + (b + c)$ is evaluated to $a + b + c$. The routine `flatten` shown below carries out this flattening for any expression with any operators. On calling flatten(e, f) the parameters e and f are evaluated first. If e is an expression and f an operator then recursively enveloped operators of the form $f(a_1..a_N, f(a_{N+1}..a_M), a_{M+1}..a_K)$ are converted to $f(a_1..a_K)$ in e. An example for this is flatten$(f(a, f(b)), f) = f(a, b)$. If e is not an expression then e is returned.

```
>> reset():

>> flatten := proc(e,f)
local do_flatten, rm_op;
begin
   # rm_op(x,f) removes the operator f in x=f(...) #
   rm_op := proc(x,f)
   begin
      if domtype(x) = DOM_EXPR then
```

```
            if op(x,0) = f then
                return(op(x))
            end_if
        end_if;
        x;
    end_proc;

    # do_flatten(e,f) -- flatten without parameter check #
    do_flatten := proc(e,f)
    local i, res;
    begin
        if domtype(e) <> DOM_EXPR then
            return(e);
        end_if;
        # flattening of operands  #
        res:= subsop(e, (i=do_flatten(op(e,i), f)) $ i=1..nops(e));
        # remove operator in the operands #
        if domtype(res) = DOM_EXPR then
            res:= subsop(%1, (i=rm_op(op(%1,i), f)) $ i=1..nops(%1))
        end_if;
        res;
    end_proc;

    # main program #
    if args(0) <> 2 then
        error("wrong no of args")
    end_if;
    if domtype(e) <> DOM_EXPR then
        return(e);
    end_if;
    do_flatten(e,f);
end_proc:

>> flatten(f(a,f(b)), f);
      f(a, b)

>> flatten(f(a,f(a),f(b,f(b))),f);
      f(a, a, b, b)
```

The attentive reader will already have noticed that the procedure **flatten** can
also be formulated in one line. Flattening is an operation that the evaluator
automatically uses with expression sequences (comma operator). If we want to
flatten with respect to an operator f then we need only to equip this with the
comma operator characteristics. For this purpose we substitute the operator f
by the comma operator **_exprseq** temporarily. By the following evaluation the

desired flattening is executed automatically. These operations are grouped in the last line of the following procedure. The other instructions are only for type testing and do not contribute to the actual functionality of **flatten2**.

```
>> flatten2 := proc(e,f)
local T;
begin
    if args(0) <> 2 then
        error("wrong no of args")
    end_if;
    if domtype(e) <> DOM_EXPR then
        return(e);
    end_if;
    f(eval(subs(op(e), f = _exprseq)));
end_proc:
```

Apart from its extremely simple structure this program has another advantage: It is noticably faster. To demonstrate this we firstly produce an expression which consists of several calls of a function f.

```
>> T := f(a),f(b),f(c):
TT := (f(T) $ 100):
```

A comparison by the **time** function impressively verifies how a carefully considered use of already existing data and structures can work on the run time:

```
>> time(flatten(f(TT),f)); time(flatten2(f(TT),f));
        26233
          800
```

3.2.8 Fast Fourier Transformation

With the next example we will show that MuPAD can also be used for numerical applications. The fast Fourier transformation plays an important role in technical applications. An iterative version of the discrete fast Fourier transformation is implemented in the program listed below. The roots of unity are described with the aid of the sine and cosine functions. A list A with complex numbers and a natural number m are expected, whereby $|A| = 2^m$ must be valid. If $A = [a_0, a_1, \ldots, a_{2^m-1}]$ is passed then these points are transformed to $B = [b_0, b_1, \ldots, b_{2^m-1}]$ and we get:

$$b_j = \sum_{0 \le k \le 2^m-1} a_k e^{2\pi I jk/N}, \quad 0 \le j \le N - 1$$

```
>> fft := proc(A, m)
local i, j, k, l, n, nd2, p2, p2m, ind, r, s, t;
begin
```

```
    n := 2 ^ m; nd2 := n div 2; j := 1;
    for i from 1 to n-1 do
       if i < j
          then t := A[j]; A[j] := A[i]; A[i] := t;
       end_if;
       k := nd2;
       while k < j do
          j := j - k; k := k div 2;
       end_while;
       j := j + k;
    end_for;
    for l from 1 to m do
       p2 := 2 ^ l; p2m := p2 div 2;
       r := 1.0; s := float(cos(PI/p2m) + I*sin(PI/p2m));
       for j from 1 to p2m do
          for i from j to n step p2 do
             ind := i + p2m; t := A[ind] * r;
             A[ind] := A[i] - t; A[i] := A[i] + t;
          end_for;
          r := r * s;
       end_for;
    end_for;
    A;
end_proc:

>> A := [2.0, 4.0, 1.0, 5.0]:

>> fft(A, 2);
     [ 1.200000000e1, 1.0 - 1.0*I, - 6.0, 1.0 + 1.0*I ]

>> A := [ 7.0, 5.0, 3.0, 6.0, 4.0, 2.0, 1.0, 8.0]:

>> fft(A, 3);
     [ 3.600000000e1, 6.535533906 + 2.707106781*I, 7.0 - 7.0*I,
       -0.5355339059 - 1.292893218*I, - 6.0, -0.5355339059 +
       1.292893218*I, 7.0 + 7.0*I, 6.535533906 - 2.707106781*I ]
```

Chapter 4

MuPAD – A Summary

4.1 The MuPAD language in brief

In this section the MuPAD language will be described — in about 11 pages. Naturally this is very short, but the most important language concepts should become clear. The evaluation of the MuPAD language shall be described in section 4.2.

Comments are one of the most important language elements. In MuPAD they are enclosed in #, as in # comment #.

Constants are numerical constants, the Boolean constants TRUE and FALSE, and character strings. Numerical constants are integers, e.g. 175, rational numbers, e.g. -17/3, floating-point numbers, e.g. 12.3e-4, and complex numbers, e.g. 2/5+4*I. (When giving exponents the user must use a small e.) Character strings are enclosed in " , e.g. "Otto".

Numerical values can have any number of digits. The precision of floating-point numbers can be predefined with the system variable DIGITS.

Special Constants are I for the imaginary unit, and E and PI for e and π respectively. The constant NIL is used to remove values from identifiers.

Identifiers are formed of an alphabetical character followed by any number of alphabetical characters or digits, e.g. deriv or x_1. (The underline _ is an alphabetical character.) An identifier can have a value assigned to it by using the assignment operator :=. At first an identifier has no value assigned to it. An identifier is of no particular type and can be assigned any value.

Function calls have the form *procedure_name (parameter_list)* or just *procedure_name()*. Here the parameter list is a series of expressions seperated by a ,. Procedure names are frequently identifiers, e.g. reset(), sin(1) or append(1,x). Procedure names can also be given in the form of expressions.

Expression sequences are series of expressions seperated by the comma operator `,`, e.g. `2, "x", y, 3/4`.

Lists are ordered expression series enclosed in `[ ]`. The expressions are seperated by `,`, e.g. `[a,b,3/7]`, `[x]` or `[]` (the empty list). The list entries can be accessed by `[ ]`, for instance, `l[3]` is the third element of list `l`. By using `append(l,7)` the 7 can be attatched to `l` and the new list is returned as the solution.

Sets are, like lists, expression series, which are enclosed in `{ }`, e.g. `{a, 2.23}`, `{"Lisbet"}` or `{}`. A set contains no identical elements and is not ordered.

Tables are any number of equations with the form `index=value`, where `index` and `value` can be any expression. Tables are produced by the function `table`, e.g. `t:=table()`. The table entries can be accessed by `[ ]`, e.g. `t[x]`, `t[7]` or `t["z"]`. (Here `x`, `7`, `"z"` are the index values of the respective expressions.)

Arrays have any number of dimensions and are used for the storing of vectors, matrices, tensors etc.. Arrays are produced by the function `array`, e.g. `v:=array(1..4)` (vector `v[1]...v[4]`), `m := array(2..6,3..4)` (matrix `m[2,3]`, `m[2,4]...m[6,4]`) or `t := array(1..3,1..3,1..3)` (a three dimensional array containing `t[1,1,1]...t[3,3,3]`). The dimensions of arrays is fixed by their declaration and cannot be changed dynamically. Not explicitly declared array elements have no values.

Polynomials are represented by a special data structure. This is generated by a call of `poly`, for instance `poly(x^3 - 2*x^2 + x - 1, [x], IntMod(7))` or `poly(x*y^3 + 2*x - z, [x,y,z], Expr)`. Admissible coefficent rings are `Expr` (expressions), `IntMod(n)` (ring of remainder classes modulo n) and Domains. More than 20 functions are available for working with polynomials.

Domains are used for the definition of new data types by the user. They are fully integrated in the kernel (so working with them is quite fast) and enable to implement polymorphic algorithms. By the function `domain`, new domains are generated, by `domattr` the methods of a domain can be accessed. Many system functions can be overloaded. For instance by defining a domain method `_plus`, the usual operator `+` can be used for addition of domain elements.

Expressions are data structures in MuPAD and not calculation statements as in most other programming languages. Expressions are not always directly evaluated. Evaluation is dependent on the context. During evaluation the identifiers are substituted by values, function calls by their solutions and arithmetic expressions are simplified and sorted for easier comparison.

Identifiers are normally recursively substituted as far as possible during the evaluation of expressions (so that in the expression no more identifiers with values are present). However, this does not apply to procedures. In procedures the identifiers are replaced by their values only once. An expression consists of a simple expression or an operator and its operands. A simple expression is a constant or an identifier, a function call, an expression sequence, a list, a set, a polynomial or a domain. Operators are MuPAD operators like `+`, `*` and `$`. The operands of an expression are themselves any expression. Examples of expressions are `a+3*PI+sin(x)` or `[f(i) $ i=1..n]`.

Tables and arrays cannot be directly included in expressions. There are only identifiers that have entire tables or arrays as values, or indexed identifiers, like `t[3]`, which have a single value as an element. Such indexed identifiers are understood internally as the operator `t` with the operand 3.

A function call, with its actual parameters, is internally understood as an operator with its operand. Conversely for every operator there is a system function which realizes the operator. An operator is executed by calling the relevant system function with the operator as the parameter.

Trees are defined by the recursive definition above. Every expression is a tree, operators and function calls are inner nodes, operands and parameter are children, simple expressions are leaves.

Arithmetical operators are `+`, `-`, `*`, `/`, and `div` and `mod`. `div` and `mod` are integer division and integer remainder respectively. `1/a mod b` results in the modular inverse of `a modulo b`.

Relational operators are `=`, `<>`, `<`, `<=`, `>`, and `>=`.

Logical operators are `and`, `or` and `not`.

Set operators are `union`, `minus` (difference) and `intersect`.

Special operators are the range operator `..`, the sequence operator `$`, the concatination operators `.` and `@`, and the domain attribute operator `::`.

An expression of type range has the form *lower_limit .. upper_limit*. The lower and upper limits are expressions that result in integers. Such a range represents the integers between the lower and upper limit.

The simplest form of a sequence is *expression* `$` *quantity*. Here an expression sequence with a *quantity* of elements is defined, with each element equal to the *expression*. The expression `a $ 3` results in the sequence `a,a,a`.

With *expression* `$` *variable* = *range* a expression sequence with the element *expression* is defined. Here, however, the running variable in the *expression* is replaced with the respective current value from the range. The expression `i*PI $ i=0..3` results in the sequence `0,PI,2*PI,3*PI`.

A sequence of integers is defined by `$` *range*. This consists of all integers between the lower and upper limits of the range. The expression `$ 2..5`

results in the sequence 2,3,4,5, $ -2..-2 in the number -2. The expression
$ 1..0 results in the *empty series*.

The concatenation operator is used to concatenate lists and character strings,
and to define new identifiers. [a,b].[c,d] produces the list [a,b,c,d],
"Can"."teen" produces the character string "Canteen". The identifier f_x
is defined by f."_x". Naturally, when defining new identifiers the result
must be a syntactically valid identifier.

The @ operator is used for composition of functional operators. For example,
(tan @ exp)(x) returns tan(exp(x)).

The :: operator is the short form of domattr, i.e. Dom::a is equivalent to
domattr(Dom, "a").

The data type of an expression is the internal type in which a expression is
stored in MuPAD. For simple expressions this is identical to the expression
type. Complex expressions have the data type DOM_EXPR. The data type can
be ascertained by the function domtype.

The expression type is the type of the corresponding object for simple expressions and for operators the operator type. Expression types are given by the
function type. type(expr) returns the type of expr, testtype(expr,typ)
results in TRUE when expr has the type typ. The expressions type(a*b),
type(x) or type(a+b+c) result in _mult (product), DOM_IDENT (identifier)
or _plus (sum). The expression testtype(x, DOM_IDENT) results in TRUE.

The Operands of an expression can be determined by the function op. op(expr)
results in a sequence containing all the operands of expr. op(expr, n)
returns the n-th operand of expr. op(expr, 0) gives the operator of expr.
If e = a+b+sin(3*c) then op(e) is the sequence a, b, sin(c*3), op(e,
1) is a and op(e, 0) is _plus.

By using op, operands can also be ascertained recursively. If e is defined as
above, then op(e, [3,1]) results in the expression c*3, op(e, [3,1,2]) is
3 and op(e, [3,0]) is sin.

Furthermore, by using op, elements of complex data structures, such as sets,
lists and tables, can be accessed. Thus, op([a,b,c], 2) results in the second
element, b, of the list.

The number of operands in an expression can be determined by the function
nops. The expression nops(a+17+3*c) results in the value 3 (for the three
operands a, 17 and 3*c).

The substitution of partial expressions can be carried out with the functions
subs, subsex und subsop. On calling subs(expr, old=new), the partial
expression old in expr is replaced by new. The partial expression old must

be complete. It is not possible to replace one operand of an operator individually and retain the others. Therefore, in the expression `e` above `3*c` can be replaced by `d` by using `subs(e, 3*c=d)`, but the partial expression `a+b` cannot be replaced by the function `subs`.

With `subsex` incomplete partial expressions can be replaced, the call is the same as that for `subs`. `subsex(e, a+b=d)` results in the expression `d+sin(3*c)`.

With `subsop(expr, n=new)` the n-th operand of `expr` is replaced by `new`. `subsop(e, 3=d)` results in the expression `a+b+c`. Internal partial expressions can be specified by giving paths instead of the simple number n (see `op`).

Statements are expressions, assignments or conditional statements. During interactive input, an statement must always be terminated by ; or : in order to be executed. When : is used the result of the statement is not shown on the screen.

Statement sequences are series of statements which are seperated by ; or :. When : is used the solution of the statement on the interactive level is not displayed.

Assignments normally bind values to identifiers, as in `a:= 12+x`. They have the form *identifier := expression*. After assigning NIL, as in `a:=NIL`, an identifier has no value.

On the left hand side of an assignment, instead of a simple identifier, an *indicated identifier*, such as `l[2]` (a list, table or array element), an identifier generated by the concatenation operator, such as `x.2`, a domain attribute, such as `Dom::a`, and a call of a function, such as `f(2)`, can be placed.

The result of an assignment is normally the right hand side of an assignment. If the assignment is NIL then the result is the identifier itself. The result of `a[3]:= 12+x*0+4;` is 16, for `b:= NIL` the result is b.

An assignment of NIL to the system variables results in them being set back to their default values. Therefore, these identifiers always have a value.

If-statements have the form

```
if condition then
    statement sequence
elif condition then
    statement sequence
...
else
    statement sequence
end_if
```

The conditions must be Boolean expressions. If the first condition results in **TRUE**, then the statements in the **then**-part are executed, if not then the conditions in the **elif**-part are examined. If these result in **TRUE**, then the statements in the **elif**-part are carried out. Any number of **elif** parts, whose conditions are examined one after the other, can follow. If none of the conditions result in **TRUE**, then the statements in the **else**-part are executed. The **elif**- and **else**-parts may be omitted. Examples are `if x > y then x else y end_if` or `if a mod 3 = 0 then a:= a+1; return(a) end_if`.

For-loops have two possible forms. In the familiar form a range of numbers is run through:

```
for identifier from beginning to end
step step width do
      statement sequence
end_for
```

The range *beginning* to *end* is run through with the step width *step width*. In each run the index *identifier* is set to the current value and the statement sequence is executed. `for i from 2 to 8 step 2 do print(i) end_for` results in the values 2, 4, 6, 8.

The step width may be omitted, in which case the default step width 1 is used. In `for i from 1 to 3 do a.i:= i end_for` the assignments `a1:=1`, `a2:=2` and `a3:=3` are executed.

The form *end* **downto** *beginning* can be used instead of *beginning* **to** *end*. In this case the loop runs backwards and the run variable is reduced by the step width each time. In `for j from 4 downto 2 do t[j]:= t[j-1] end_for` the assignments `t[4]:= t[3]`, `t[3]:= t[2]` and `t[2]:= t[1]` are executed in the given order.

In the second form of the **for**-loop all operands of an expression are run through:

```
for identifier in expression do
      statement sequence
end_for
```

The operands of the expression are assigned, one after another, to the identifier and then the loop body is executed each time. In `for t in a+b+c do print(t) end_for` the values for `a`, `b` and `c` are displayed one after the other.

While-Loops have the form

```
while condition do
    statement sequence
end_while
```

Here, the statement sequence is executed as long as the condition results in TRUE. The condition is evaluated before the statement sequence is executed.

Repeat-Loops have the form

```
repeat
    statement sequence
until condition end_repeat
```

Here, the statement sequence is executed as long as the condition results in FALSE. The condition is checked after the evaluation of the statement sequence.

Any loop can be terminated by **next** and **break**. With **next** the execution of an statement sequence is terminated and the next run of the loop is started, with **break** the loop is terminated.

Case-statements have the form

```
case expression
    of value1 do statement sequence
    of value2 do statement sequence
    ...
    otherwise statement sequence
end_case
```

Here, the expression is compared with the expressions *value1*, *value2*, etc. one after the other. If the expressions match, then all the following statement sequences of the **case**-statement are executed. Also the sequences of the remaining **of**- and **otherwise**-parts! If the expression does not match to any of the **of**-parts, then only the statements of the **otherwise**-part are executed. The **otherwise**-part may be omitted.

With the **break** command the user can jump to the end of a **case**-statement, the rest of the statements in the **case**-statement are not executed. With the **next** command the comparison of the expression with the values of the **of**-part is resumed: Like at the beginning of the **case**-statement, the comparison of the expression with the values of the **of**-parts is resumed starting with the **next**-statement and the statements are executed after an appropriate value is found.

Procedures have the form

```
proc ( parameters )
    local local variables ;
    option options ;
    name name ;
begin
    statement sequence
end_proc
```

The formal parameters are a (possibly empty) series of identifiers seperated
by a ,. The local variables are also a sequence of identifiers seperated by a
,. Options are keywords seperated by a ,, **hold** and **remember** are accepted.
The name can be any datum and is used only for output. I.e. the procedure
is not entered in the stack as a value for the name. The declaration of
the local variables, options and name may be omitted. (Sometimes it is
advantageous to use procedures without names.) If the specification of the
name is missing, then with the assignment of a procedure to an identifier,
the identifier is taken as procedure name. With

```
max:= proc(x,y)
     begin
       if x > y then
         x
       else y
       end_if
     end_proc
```

the procedure `max` is defined, this calculates the maximum of two numbers.
With the assignment `f:= max`, `f` is assigned the same procedure as `max`. In
MuPAD, procedures are data like expressions and lists.

On calling a procedure, the formal parameters are assigned copies of the
actual parameters. (A "*call by value*" is executed, a "call by reference" does
not exist.) Then the body of the procedure is executed. At the beginning
local variables are unassigned.

A *free variable* in a procedure (a variable that is neither a parameter nor a
local variable) is automatically a global variable. A procedure can be defined
in another procedure. In this case the local variables of the outer procedure
can be global variables of the inner one, as in

```
g:= proc(x)
     local y, f;
    begin y:= x;
      f:= proc(x)
```

```
            begin
              x*y
            end_proc;
          f(x)
        end_proc
```

Here the inner procedure **f** is a local variable of the outer procedure **g**, **y** is a local variable of **g** and a global variable of **f**. **g(x)** evaluates (very clumsily) to x^2.

Dynamic scoping is used for the free variables, i.e. global variables are bound during runtime. (In a procedure, those global variables which were last defined in one of the procedures called so far, are used.)

Normally, a procedure returns the value of the previously executed statement. A procedure can also be left by using the function **return**. The parameters of **return** are the return values of the procedure. The above mentioned procedure **max** can also be defined as follows:

```
max:= proc(x,y)
      begin
        if x > y then
          return(x)
        end_if;
        y
      end_proc
```

The return value of a procedure can be an expression sequence, with this a procedure can return several values at the same time.

Options for procedures are, as already mentioned, **hold** und **remember**. Normally, the current procedure parameters are evaluated before the procedure is executed. This mechanism is prevented with the option **hold**. The option **remember** causes a procedure to "remember" the return values and if necessary simply to return these values instead of executing the procedure again. By this means especially recursions can be greatly accelerated at the cost of using more memory.

Environment variables are system variables that are automatically re-set to their old values after a procedure has been executed. The procedure

```
pi:= proc()
       local DIGITS;
     begin
       DIGITS:=200;
       float(PI)
     end_proc
```

returnst π to exactly 200 digits. After calling `pi()`, the system variable `DIGITS` still has the same value as before the call, i.e. it was only altered during the procedure `pi`. However, for this the environment variable must be declared as `local`!

Parallel statements are simple to use in the area of *micro-parallelism*, and there they offer an optimal adaptation to the basic hardware. Micro- parallelism can be employed efficiently when communication between the processes involved is minimal. The system carries out *load balancing* and *process communication* independently. There are only the parallel implementable program parts to mark. The constructs of *macro-parallelism* provide a tool for optimal use of the hardware topology. Load balancing and process communication are no longer carried out automatically but have to be done by the user.

The micro-parallelism is based on the model of the shared memory machine. With the available constructs processes are generated and deposited on a *task stack*. Idle processors examine this stack and remove tasks for processing. There are a parallel for-loop and a parallel block, both of which are also available in the sequential version of MuPAD.

```
for identifier from beginning to end step step width parallel
    private private variables ;
    statement sequence
end_for
```

The range from *beginning* to *end* is run through with the *step width*. In each run the index *identifier* is set to the current value and a process is generated by the *statement sequence*, which is then deposited on the stack. The private variables are a (possibly empty) series of identifiers which are seperated by a `,`. The usage of these variables is analogous to local variables in procedures.

A second form of the parallel for-loop runs through all operands of an expression:

```
for identifier in expression parallel
    private private variables ;
    statement sequence
end_for
```

As in the sequential case, all expression operands are assigned, one after the other, to the identifier. For each value a process is deposited on the task stack.

A parallel loop is only then ended when all the tasks it has generated are executed. The return value is a expression sequence containing the values of each executed task. The result of `for i in [1,2,3,4] parallel i^2 end_for` is 1,4,9,16.

With the help of parallel blocks any statement or statement sequence can be run in parallel.

`parbegin` `private` *private variables* ; *statement sequence* `end_par`

Each statement in the *statement sequence* is deposited as a task on the task stack. In order to execute statement blocks sequentially inside a parallel block, a sequential block is available:

`seqbegin` *statement sequence* `end_seq`

Macro-parallelism can be efficiently used in distributed systems, because, in this case, all communication has to be explicitly defined by the user. Here, the system does not care for any load balancing or scheduling tasks. As a model a network of *clusters*, which are independent of one another, is used. These clusters can communicate with each other by exchanging messages.

Queues have names which can be any MuPAD datum. They are used to enable the communication of clusters with each other. A queue is organized on the *first in first out* principle. With `writequeue(name, i, value)`, the `value` is written in the queue called `name` of the cluster i. With `readqueue(name)`, a cluster reads the first element in its queue. As special queues those with the name `"work"` are available on every cluster. If a cluster is idle, it tries to read a datum from this queue and evaluate it.

Pipes allow another form of communication between clusters. Because two clusters can both write into the queue of a third cluster, it is possible that the receiver has problems to identify the sender of the last message. Pipes offer the possibility of building selective two-way communication. With `writepipe(name, i, value)`, the `value` is written in the pipe called `name` of the cluster i. Through `readpipe(name, i)`, a cluster reads the next message in the pipe `name` of the cluster i.

Net-variables constitute a third form of communication possibility in macro-parallelism. An identifier a, which is assigned the value b by `global(a,b)`, has this value in all clusters. Therefore, net variables constitute a distributed shared memory.

4.2   Evaluation in MuPAD

In MuPAD terminology, evaluation is the execution of a MuPAD expression by the MuPAD interpreter. It can be said, with good reason, that evaluation is the "actual purpose" of the interpreter. In this context, what does "execution" mean?

4.2.1   MuPAD expressions

Before evaluation is tackled, the term expression must be explained: A MuPAD
expression is — expressed very simply — a tree.

The leaves of an "expression-tree" are basically identifiers (i.e. variable identifiers),
constants (numbers, Boolean constants, strings) and special constructs, that are
needed to identify system functions and procedures.

Identifiers have, apart from the fact that they have names, another extremely
useful characteristic: Values can be *bound* to identifiers i.e. an expression can be
allocated to an identifier. (This is done with the assignment ":=".) Through
evaluation, an identifier, which is bound to a value, is substituted by this value
(i.e. by this tree). For instance, with the assignment

```
a:=b;  b:=2;
```

the value b is bound to the variable a and the value 2 to b.

The inner nodes of a tree are function calls or complex data structures, like lists,
sets or tables. In the following, complex data structures will not be dealt with, in
order not to make the description too complicated.

With function calls the first subtree identifies the function and the other subtrees
identify the current parameters of the call. Such a node can be written as

$$F(P_1, \ldots, P_n)$$

in "functional notation", where the function F and the parameter P_i can stand for
subtrees (not only for leaves). In Lisp these nodes would be written as $(FP_1...P_n)$.

The statements of the MuPAD language are all functions and are also accessible
to the user as functions (e.g. as **_assign** for the assignment ":=" or as **_if** for
the "**if then else end_if**" statement). For each MuPAD statement there is a
so called *"underline"-function* (so called, because by convention all the names of
these functions begin with an underline "_"). The MuPAD statement

```
if a < b then a+2 else b end_if
```

can be just as well be written as

```
_if(_less(a,b), _plus(a,2), b)
```

Here the legibility obviously suffers.

4.2.2 Evaluation of expressions

Through *evaluation* a MuPAD expression is transformed into another one, in other words one tree is replaced by another. In MuPAD evaluation is a recursive process:

A leaf, which is not an identifier, is not altered through evaluation. Likewise, if an identifier has no value bound to it, then it is also not altered. When an identifier has a value bound to it, then this value (i.e. this expression tree) is evaluated and the solution returned. Following the statement

```
a:=b; b:=c; c:=2;
```

a is normally, during interactive input, evaluated to 2. (More to the qualifications "interactive" and "normally" later.)

As already mentioned the inner node of an expression is a function call. Initially the first subtree of the node is evaluated. If the evaluation of the first subtree results in a function definition then the corresponding function is carried out with the remaining subtrees of the node as parameters. During the evaluation of the call

```
sin(a+b+2)
```

the identifier sin is evaluated to the function definition of the system function "sine" and the function can then be called with the expression a+b+2 (or the equivalent _plus(a,b,2)) as parameter.

If the evaluation of the first subtree of a function call results in an identifier, then the function is not bound to a definition. In this case, all the remaining subtrees are also evaluated and the new tree is returned in the form of a "formal function call". If f is, for instance, an identifier to which no value is bound, then

```
f(4+3)
```

will be evaluated to f(7) because _plus(4,3) will be evaluated to 7.

If the first subtree of a node is evaluated neither to a function definition nor an identifier then the evaluation is terminated with an error. Therefore the expression

```
"_plus"(2,3)
```

is invalid, because the string "_plus" cannot be evaluated to the function definition for the sum.

4.2.3   Execution of MuPAD-Procedures

As already mentioned, a function is executed with the subtrees of the call as
parameters. If the function is a procedure which is defined in the MuPAD language,
then there are two forms of call: Normally, the parameter expressions are evaluated
and the procedure is executed with these parameters. The statements

```
f:= proc(x) begin
   x*2
end_proc;
a:=2;
f(a);
```

define the function f and then assign the value 2 to a. As expected the call f(a)
returns the value 4.

Conversely, if in the procedure definition the option "hold"is given, then the parameters are not evaluated before execution. The statements

```
g:= proc(x) option hold; begin
   x*2
end_proc;
a:=2;
g(a);
```

define the function g and again assign the value 2 to a. The call g(a) then returns
the expression a*2 as the solution: The parameter a is not evaluated, the function
g returns the identifier a multiplied by 2 as the result.

The execution of the procedure takes place in three steps: Firstly the actual parameters are bound to the identifiers of the formal parameters. Then the local
variables of the procedure which have been declared "local" are established. In
the last step the procedure is evaluated, as already described. The final evaluated
value is then the return value of the call.

The local variables are virtually "fresh" identifiers, to which, at first, no value is
bound. If identifiers with the same name exist before the procedure call, then
these values are not visible during the procedure. Instead, during the evaluation
of the identifiers the local variables are used. All other identifiers are handled as
global variables; in this context the current, valid values of the identifier for this
run time are used.

In the following procedure f, for instance, x is the parameter and y is a local
variable, while z is a global variable, because the identifier z is declared neither as
a parameter nor a local variable:

```
f:= proc(x) local y; begin
   y:=2; x*y*z
end_proc;
```

After the assignments

```
y:=3; z:=NIL;
```

the value 3 is bound to y, while z (due to the assignment z:=NIL) has no value. Because in the expression x*y*z the local variable y is evaluated to 2, afterwards the expression f(2) gives the value z*4. However, after the assignment

```
z:=3;
```

f(2) returns 12.

The described method has one consequence, this is called *dynamic scoping*: For global variables those identifiers are used which are visible at the *execution time*. In a "normal" procedural programming language, like Pascal, those variables are used which are statically declared through the program text (through the outer procedure). This is called *lexical scoping*. In the following procedure g, the local variable z, which was global in connection with the function f above, is defined:

```
g:= proc() local z; begin
    z:= 4; f(2)
end_proc;
```

Here, the expression g() always returns the value 16. In this context, the value of z makes no difference to the call of g. During the evaluation of f(2) in the procedure g, the local variable z from g is used (this was not "known" in the definition of f).

4.2.4   Execution of System Functions

In the system functions, parameter evaluation is dependent on the given function. Most functions evaluate all their parameters before the execution of the "actual" function. However, some functions do not, or only partially, evaluate their parameters.

The assignment (**_assign**), for instance, does not evaluate its first parameter (the right hand side of the assignment), if this is an identifier. Otherwise the variable b in the assignment

```
a:=b; a:=2;
```

would have the value 2 — an extremely confusing situation. However, the left hand side of the assignment is evaluated. Through the assignment

```
b:=2; a:=b;
```

the value 2, and not the value b, is bound to a during interactive input.

Another example for incomplete evaluation of parameters is the procedure definition with `proc ... end_proc`. In this case the procedure definition is not evaluated. It would be very confusing if, with the assignment

```
x:=13;
f:= proc(x) begin x*2 end_proc;
```

the (incorrect) procedure definition

```
proc(13) begin 13*2 end_proc;
```

was generated.

Of course, the execution of system functions is dependent on the given function. In the following, only a few directions for the execution of functions, like "+", "*" und "^" which represent the algebraic operations, are given:

With these functions the evaluated parameters are at first simplified and then transformed into standard form, in order that the expression can be more easily compared with other expressions.

Simplification is carried out with simple rules. Firstly the associativity of the operation is used, in order to "flatten" the expression tree. Encapsulated expressions of the form

```
_plus(a,_plus(b,3))
```

can be simplified to

```
_plus(a,b,3)
```

due to the associativity of addition.

After this rules like $a + 0 = a$ or $a \times 0 = 0$ are used. As a consequence of this all 0's are removed from sums and multiplication with 0 results in 0. The use of these rules can be — and this drawback cannot be denied — quite hazardous. One such hazardous rule is $a/a = 1$; because it is possible that later a can take the value 0, so the substitution with 1 is wrong in this case. If the variable a is defined by

```
a:= b*(c+1)/b;
```

with b not yet having a value, then a is given the value c+1 through simplification. If afterwards

```
b:=0
```

is defined, then the value of a is still falsely defined as c+1, the variable a does not "know"the variable b from its definition any more.

Unfortunately, here, the user must live with this hazard, as otherwise expressions can very quickly "explode". The user should be aware of the hazards of the simplification rules.

The calculation of constants also belongs to the simplification rules. This includes transformations like

```
_plus(3,a,4)  ⟶  _plus(a,7)
```

After simplification the expression is normalized. Here, for instance, in sums and products constants are moved to the end of an expression and the remaining expressions are sorted. The expression

```
z*3*a1*a
```

is transformed into

```
a*z*a1*3
```

by sorting. The sorting order is difficult for the user to predict, but luckily this is nearly always irrelevant.

The expression b*a is always transformed to a*b through sorting. Therefore, sums and products must always be conceived as commutative. If matrices or non-commutative operators are to be multiplied using symbols then the user will need to use his own functions.

The calculated expression is then returned as the solution of the evaluation.

4.2.5 The Substitution Depth

In the evaluation of identifiers the *substitution depth* plays an important role. What is to be understood by this?

If the original expression is considered before evaluation then the substitution depth is 0. Each time an identifier is substituted by its value the substitution depth is increased by 1 and the value of the identifier is evaluated with this new substitution depth. After evaluating the value the old sustitution depth is taken on again.

This means that the substitution depth gives the recursive depth of the evaluation of identifiers. In this case evaluation can be controlled by declaring a maximal substitution depth. If the maximal substitution depth is reached then the identifier is not evaluated any further.

Interactively entered expressions are fully evaluated (fully means the user can still give a maximal substitution depth). However, normally the expression is evaluated to the point where no unbound identifiers are present. After the assignment

```
a:=b; b:=c; c:=13;
```

the variable a, when interactively entered, is evaluated to 13, as expected.

In MuPAD procedures the maximal substitution depth is 1, i.e. identifiers are substituted by their values, these are, however, not evaluated further. In the following procedure f the local variable z is evaluated to y and *not* to the parameter x:

```
f:= proc(x) local y, z; begin
   z:=y; y:=x; z
end_proc;
```

The call f(13) returns the solution y and not 13, as might have been expected.

4.2.6   Control of the Substitution Depth

The user can explicitly set the maximal substitution depth with the variables LEVEL and MAXLEVEL.

During evaluation substitution is carried out (when neccessary) to the depth LEVEL. With the assignment

```
a:=b; b:=c; c:=13;
```

a is normally evaluated to 13 in interactive input. If however LEVEL is set to 2 then a is evaluated to c and b to 13.

By the way, LEVEL can be set to values other than 1 during procedures. In the procedure

```
f:= proc(x) local y, z; begin
   LEVEL:=2; z:=y; y:=x; z
end_proc;
```

z is evaluated to the parameter x and not to y. The call f(13) results in 13 and not y.

The variable MAXLEVEL is used to detect infinite recursions, as in the seemingly harmless statement sequence

```
a:= a+1; a;
```

If a has no value before the assignment, then a is evaluated to a+1 in the second statement, this then to a+2, this to a+3 and so on, ad infinitum.

In order to detect such situations, a substitution depth of MAXLEVEL, during evaluation, is taken as an error and evaluation is terminated with the message "Recursive Definition". Naturally this is only a heuristic, which may lead to false error messages in certain cases, but in most cases it is correct.

4.2.7 Fine Control of Evaluation

So far, for controlling evaluation, the procedure option `hold` and the variables
`LEVEL` and `MAXLEVEL` have been described. The system functions `hold`, `val`, `level`
and `context` allow even greater control of evaluation.

The function `hold` can have any expression as a parameter. This is not evaluated but returned as the solution of the call. So, with `hold` the evaluation of an expression is prevented. For instance, the expression

```
hold(a+2+0)
```

is evaluated to

```
a+2+0
```

and not simplified any further.

The function `val` also has an expression as a parameter, which, exactly like `hold`
is not evaluated. Furthermore all identifiers in the expression which are bound to
a value, are substituted with this. This expression is then evaluated no further,
but immediately returned as the solution of `val`. For instance after the assignment

```
a:=b; b:=2;
```

the expression

```
val(a+3+0)
```

is evaluated to `b+3+0` and the expression

```
val(b+4)
```

to `2+4`, but both are simplified no further. The most important use of `val` is
the quick and direct access to the value of a variable, without a time consuming
evaluation of the value taking place.

The function `level` is given an expression as its first parameter and, optionally,
an integer as its second parameter. An expression with the form `level`(a,n) is
evaluated, by the expression a being evaluated to the maximal substitution depth
n instead of to the "current" depth given by `LEVEL`. The function `level`, therefore, temporarily overwrites the variable `LEVEL`. The variable `MAXLEVEL` remains
unchanged, on reaching the substitution depth `MAXLEVEL` an error message will
still be given. If the second parameter is missing then a nearly infinite maximal
substitution depth is assumed (to be exact 2147483647).

On the interactive level, with the statement

```
a:=b; b:=c; c:=3;
```

the expression

```
level(a,1)
```

is evaluated to b and the expression

```
level(a,2)
```

to c. Also, the function `level` can be especially used to fully evaluate a variable in a function.

`context` can be used for evaluation of arbitrary expressions in an "outer context". If `context` is called in a procedure, then the argument, which can be of any type, is firstly evaluated in the actual context. The result is then evaluated in the context, from where the procedure has been called. For instance:

```
i := 2:  a := "a":  b := "b":
f := proc(x)
option hold;
local a,b;
begin
   a := b; b := x;
   print(x, context(x));
   print(a, context(a));
end_proc:
```

By calling

```
f(i)
```

one gets the result

```
i, 2
b, "b"
```

Chapter 5

Tools and User Interfaces

5.1 Graphical User Interfaces under X-Windows

MuPAD is written in the programming language C. The executable code takes up less than 1 MB space on a Sun Workstation. Constructed on a memory management module, the kernel is the "heart" of the system. Here all the users input is read and processed. In order to work with MuPAD no special input or output devices are needed. A keyboard and ASCII terminal are all that are neccessary. If the user has a Sun SPARCstation with X-Windows or OpenWindows to work with, then MuPAD is even easier to use. Under these window systems, the user-friendly interfaces make working with MuPAD easier. These are XMuPAD, the front-end of the actual kernel, the front-end mdx of the MuPAD debugger and the interactive graphic tool VCam. This package is rounded off by the hypertext system HyTeX, with which the complete MuPAD documentation, including the on-line help, is made available on the screen.

5.1.1 XMuPAD

XMuPAD is a fully independent program. It has no knowledge of computer algebra. It is implemented as an interface to the MuPAD kernel with the help of the XView-Library. XMuPAD is invoked with the command

```
xmupad [options]
```

The only important option for the user is the -L option. The language of the on-line documentation can be stipulated with this option. Possible parameters are german or english.

After being called up a *Basic Window* (see figure 5.1) appears on the screen, which consists of two components:

- a row of buttons and

- a text window for input and output.

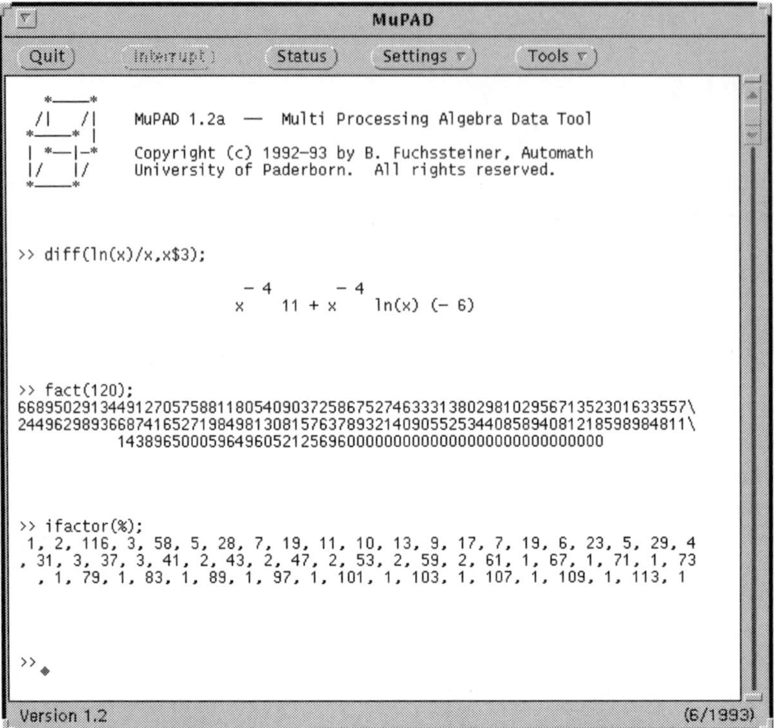

Figure 5.1: XMuPAD-Basic Window

Furthermore the MuPAD kernel is invoked as a son process on the UNIX level.

XMuPAD's function can be divided into three phases, which are typical for interactive systems:

1. During the *input-phase* the user enters commands into the text window. This phase is started with the output of the MuPAD prompt character (>>). It is ended with the <Return> key.

2. During the *evaluation-phase* the input of the first phase is evaluated.

3. The third and final phase is the output of the solutions MuPAD has calculated.

Input and output of commands takes place in the text window. In this the user has a simple, mouse sensitive text editor at his disposal. By pressing <Return> the previously entered text is read and passed to the kernel for evaluation. XMuPAD independently detects which text belongs to the current input. This can be characters before or after the current cursor position. If the input line is ended with <Enter> then a new line is started at this position, however, the entered line is not passed to MuPAD for evaluation. In this way procedures or instruction blocks can be entered.

By simply shifting the cursor to already present input and pressing <Return> again this input is read and re-evaluated. If, in the mean time, identifiers have been given new values, then these are taken into account during execution. In the settings menu it can be determined if the old output is to be overwritten or if the new output should be inserted (replace mode). In this menu it can also be determined if jumping back to old input results only in a new evaluation of this input, or the evaluation of all following input (recalculate mode). This offers a very easy method of repeatedly carrying out an interactive entered sequence of instructions, if neccessary, with altered starting values.

The functions of XMuPAD and MuPAD differ in a few points. These concern the execution of the system functions `system`, `textinput` and `input`. With `system` a new window is opened, in which a UNIX-Shell is made available and the parameter of the `system` command is executed. In MuPAD the user gets no information whether the command has been executed successfully or not. Thus, `system` is only suitable for interactive use. With `textinput` and `input` further windows, in which the desired text can be entered, are opened.

5.1.2 HyTeX

HyTeX is a hypertext system based on the TeX system. It is used for the complete on-line documentation. This includes the MuPAD reference manual and on-line help, with help pages for each MuPAD command. HyTeX is started together with XMuPAD as a son process but at first is only visible as an icon. Only on an explicit inquiry from the user, by entering the appropriate help command, is HyTeX opened.

5.1.2.1 HyTeX-Window

The HyTeX window (see figure 5.2) contains a LaTeX previewer, which is provided with hypertext functions. The previewer gives various navigational possibilities through the documentation.

Apart from the possibility of turning the pages forwards and backwards there is a so called "Page-Slider" which enables the user to choose and jump to any page of the document. But the actual highlights are the "Hyper-Links". In the text some

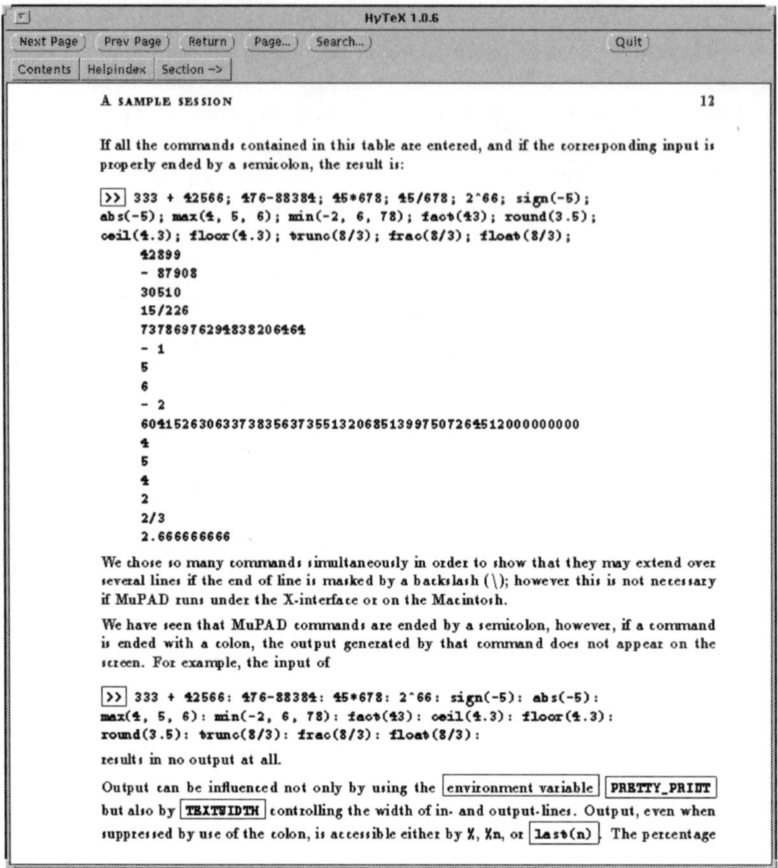

Figure 5.2: HyTEX-Window

words are framed with a box. With this references to other passages in the text, also called nodes, are marked. Generally, these contain further information to the thus framed text. By clicking such a box, the user jumps to the corresponding node. In this way, the user has the possibility, with individually chosen cross-references, to "skip" through the document. A second form of Hyper-Link is to be found in the top line of the HyTEX window. With these buttons the user can jump to the next or previous chapter, paragraph, etc..

If the user has made such a jump he will naturally want to return as quickly as possible to his starting page after reading the extra information. For this purpose there is the return button, with which the user can return to the page he was on before the "Hyper-jump".

Many examples in the on-line documentation and tutorial are provided with a button. By clicking these the user can copy the example input into the XMuPAD window. Now the user can study this command once again, make any desired changes and by pressing <Return>, it can be executed. In this way the user can easily get to know MuPAD on the screen, without previous knowledge of the MuPAD data structures and the MuPAD language syntax.

5.1.2.2 Help-Pages

As already mentioned above, beside the MuPAD-reference manual, the on-line help system is also integrated in HyTeX. For example, in MuPAD or XMuPAD the user can request information about the system function `igcd()` with the command

```
help("igcd");
```

or the short form

```
?igcd
```

Then, in HyTeX the corresponding page is opened (see figure 5.3). The exact command syntax is described on such a help page, a short summary and some examples for use are given, and jumps to the help pages of related functions or commands are possible. Here, each example has also been given a button. Analogue to text, the example is transfered to XMuPAD and written in the XMuPAD text window. Now the user can study the example at his leisure, modify it if neccessary and then carry it out by pressing <Return>. With this "Learning by doing" the user quickly gains knowledge of the syntax to be used.

5.1.3 Debugging in MuPAD - mdx

MuPAD contains two possibilities of following the program run in detail. The *trace* mode makes a protocol of the program run. In this mode the user has no influence on the program execution. This possibility is, however, offered by the interactive, line-oriented source code debugger, which has already been introduced in the section 2.14. By running through the program step-by-step and setting breakpoints the user has exact control over the execution of the program. Furthermore by displaying and altering identifiers the user also has the possibility of influencing the execution of the program. This debug mechanismus, which is integrated in the MuPAD kernel, has a character/line oriented user interface and offers, even on a text-only output device (ASCII terminal), operational convenience and very detailed information in text form. An interface based on OpenWindows, which offers even more comfort, is the program mdx. This is not integrated in the MuPAD kernel but it is a separate tool. It enhances the operational convenience by:

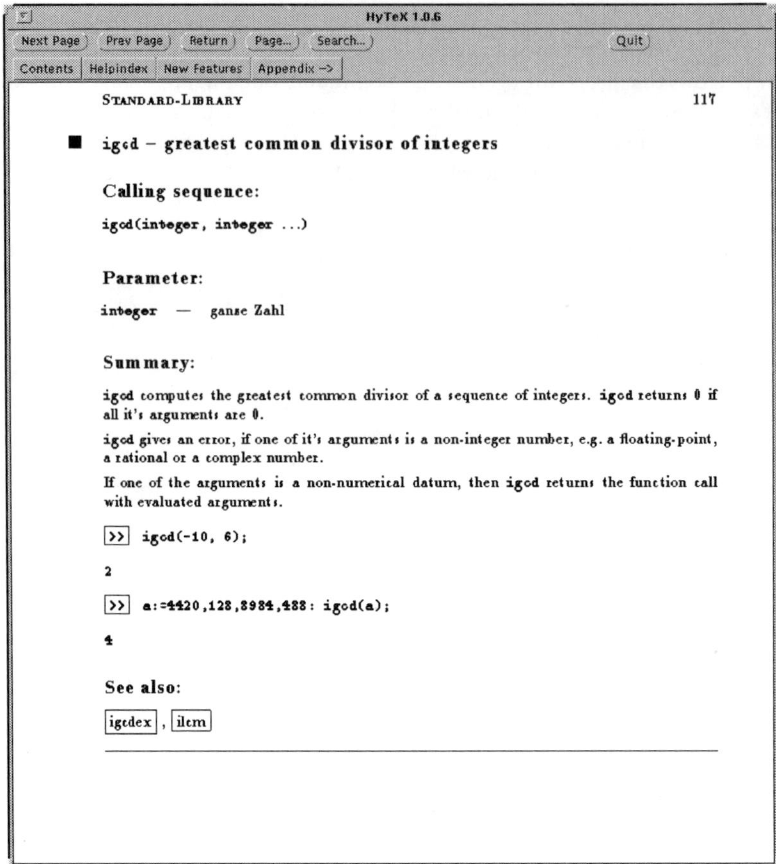

Figure 5.3: Help-Page for `igcd`

- Simplifying the input of debug commands

- Visually processing the information provided by the debugger

- Simultaneously displaying the source text and program output in separate windows

With the command

$$\texttt{mdx [options]}\ \mathit{file}$$

the debugger front-end is started. The options -H *size1, size2, size3* are available. With *size1, size2, size3* the heights (in pixel) of the three windows; *source text, command* and *display*, are given. A good setting is the choice -H 300, 200, 150. If a file *file* is transfered this is interpreted as MuPAD program code and evaluated. The front-end can be seen in figure 5.4. It consists of five windows:

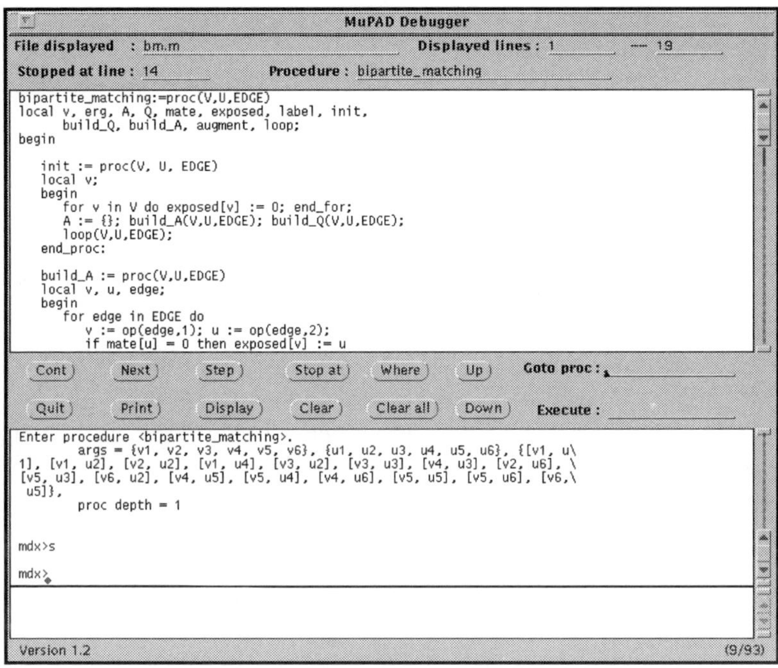

Figure 5.4: mdx-Window

Status presents general information about the current state. This contains, among other things, the line number, in which the program was stopped, and the file which is displayed in the source-text window.

Source-text shows the file which is currently being processed. If a file was specified in the call then that file is displayed in this window. Otherwise, this window remains empty until a **read** command is executed. In the course of the program execution, the text passage which contains the line in which the program was stopped, is shown. Which text passage the window should display the user can also explicitly determine with the goto proc button.

Buttons Debugger commands which are often used can also be executed by clicking with the mouse. Commands which are not included can be entered directly into the command window with the keyboard.

Command All output produced by the execution of a MuPAD program, and some output evoked by debugger commands, is displayed in this window. Both MuPAD and debugger input must be entered here.

Display The output of the variable values, which need to be permanently shown, takes place in this window.

The source text, command and display windows all have a scrollbar on the right hand side. This is to enable the user to see the entire text in each of these windows.

Naturally, in the X front-end, the identical commands, which can be used in the terminal version of the debugger, are available. For most of the possible commands there is a button or a special input panel. Of course, the user can also use the terminal versions short form for the command in the command window. In the following we would like to briefly present the commands.

Next Short form: n

Execution of all instructions in the current line. The debugger then stops at the line that is next to be executed. If one of the executed instructions of a call contains a user defined function then this function is carried out without the debugger stopping in it.

Step Short form: s

Analogue to next, only here the debugger stops in a user defined function.

Cont Short form: c

The program is carried out until it finds an instruction in a line that has been given a breakpoint or until the program is completed. See also Stop at.

If the debugger stops then the line in which it stopped (`Stopped at:`), in which file this line is (`Displayed file:`) and in which procedure it has stopped (`Procedure:`) are shown in the status window. The source text window contains the file passage in which the debugger stopped. Which lines are contained in the window are given by `Lines:` ... -- .... The current line is also inversely shown.

Goto proc Short form: g *name*

The specified procedure in the input field or the actual procedure is displayed in the source text window.

Quit Short form: q

Terminates the execution of the MuPAD program, when the debugger is in input mode. The debugger then reports `Execution completed`, and the user is in the MuPAD input mode. Quit in the MuPAD input mode terminates the debugger. An already running program can be halted with <Ctrl-C>. The debugger is then in input mode.

Execute Short form: **e** *command*

> The specified MuPAD command is executed. The command need not only contain single instructions, but can be made up of any number of instruction sequences. One use is the interactive influencing of the program run by changing the values of some variables.

Where Short form: **w**

> All the procedures which were called between the beginning of the program run and this point in time, are displayed in the terminal window. Apart from the procedure names the user is also given information about the call position in the form of a line number and a file name.

Up Short form: **u**

> In the source text window that line is displayed, from which the procedure shown so far was called. The debugger status lines are also brought up-to-date. If the procedure that the user interactively entered has been reached, then the message **Top level reached** is shown in the command window.

Down Short form: **d**

> This can only be used when at least one up has occured. The procedure is shown in the source text window from which the up was called. If the current procedure, i.e. the procedure in which the debugger has stopped the program run, has been reached then the message **Bottom level reached** appears in the command window.

Clear all Short form: **a**

> Deletes all breakpoints.

Clear Short form: **C** *filename line*

> The user specifies the line in which the breakpoint is to be deleted with a double click of the mouse button.

Stop at Short form: **S** *filename line*

> The user specifies the line in which the breakpoint is to be entered with a double click of the mouse button. In order to enter a breakpoint in a specific procedure this must already have been loaded in the source text window. See also Goto proc.

Print Short form: **p** *expr_1 ... expr_n*

> The user specifies the expression whose current value is to be shown by marking the expression in the source text window with the mouse. The value of the variable is then shown in the command window.

Display Short form: D *name_1 .. name_n*

> The user specifies the variable whose value is to be shown by marking the variable name in the source text window with the mouse. The value of the variable is then shown in the display window. From this time on the current value of the variable is permanently shown. However updating only takes place when the debugger is in input mode, i.e. when the program run is stopped.

List Short form: l

> Lists all breakpoints.

Each of the previously described commands, which can be executed by clicking with the mouse on the appropriate button, can be directly entered into the command window with the keyboard.

5.1.4   VCam - The interactive MuPAD graphics

Exactly like the other user interfaces, the front-end VCam of the MuPAD graphics is an independent tool. On starting XMuPAD it is also loaded but not yet visible. On entering the command

> `plot2d`(*parameter*) or `plot3d`(*parameter*)

the given parameters are evaluated, the graphic data calculated and then transfered to VCam to be depicted on the screen. For the calculation of graphics it is necessary to enter the graphic object in parameter form even if this curve or surface can be explicitly written down (see section 2.13).

The production of graphics is based on the photography process. To get a better understanding the user should imagine that he wants to take a photo containing specific information. Then objects with special features are grouped into a scene together with other characteristics, which are valid for all objects. This scene is now "photographed". The user must determine which parts are to be displayed (*Zooming*), which measurements fixed (*Scaling*) and how these are to be marked (*Axes, Ticks, Labels*), where the camera is positioned (*Perspective*) and which lighting (*Lighting*) is to be used in the scene. The object features are determined by the title, the coordinate functions, the parameter range, the plot style and the colors. All these options can be given as arguments in a plot command. However, it is much easier to set the characteristics interactively in VCam. Besides the base window (see figure 5.5), in which the graphic is plotted, there are graphics manipulation windows for two and three dimensional scenes. With the aid of these windows, the user is able to manipulate existing graphics and on the other hand, can interactively generate new graphics. For the latter, a call of **plot2d** or **plot3d**

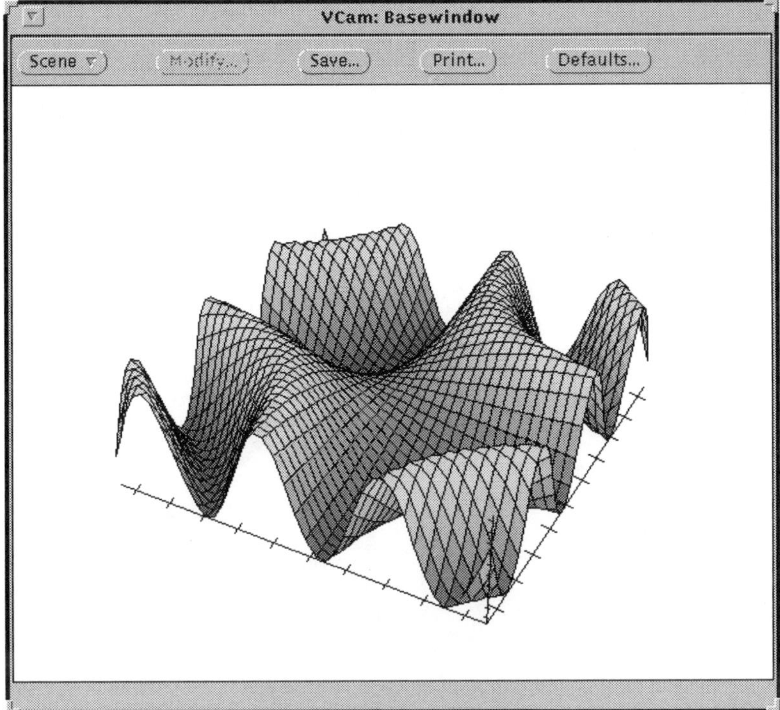

Figure 5.5: VCam-Base Window

without arguments or the selection of Graphics in the XMuPAD tools menu is the appropriate method. These open the base window with an empty drawing area.

A menu, in which the user can choose between create 2D-scene and create 3D-scene, is concealed behind the button scene. If the user chooses the latter a graphics manipulation window for a three dimensional scene is opened (see figure 5.6). All of the previously listed characteristics can be set here. In the upper part of the window the scene-specific characteristics are set, in the lower (larger) part the characteristics of the objects are shown and can be chosen. The desired settings can be entered in the window and by activating the plot button a `plot3d` command with the relative parameters is produced. This command is transfered to MuPAD for evaluation, the graphics data are calculated and finally drawn in the drawing area of the basic window. Hence it is possible to start new graphic commands (and with this new MuPAD processes) from VCam.

To illustrate this interactive procedure we shall construct an example graphic, in which we shall make the neccessary entries step by step. At the same time we shall introduce some of the many possible settings. We have given ourselves the

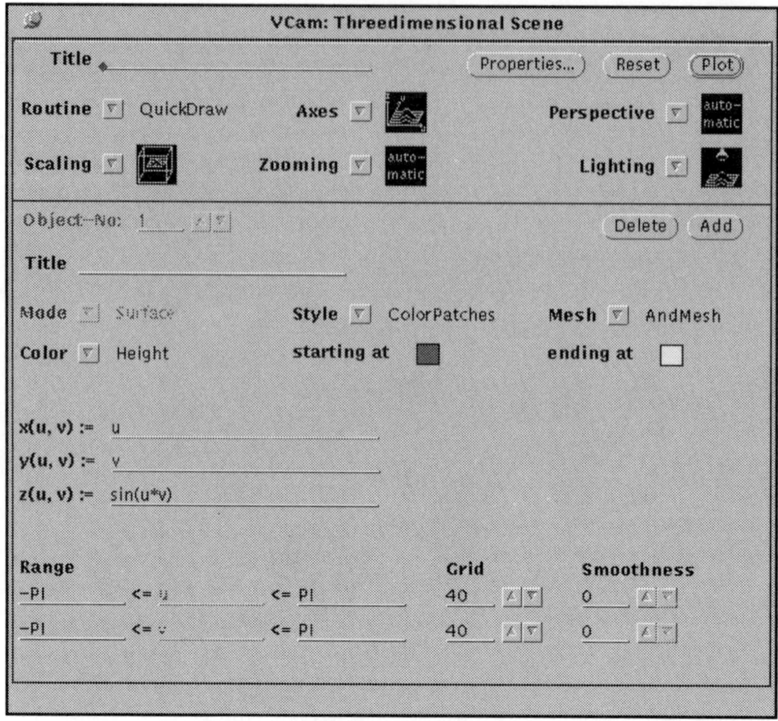

Figure 5.6: Graphic Manipulation Window for a 3D-Scene

task of drawing a scene which consists of a sphere. We need to start with the object specific characteristics. Firstly, we add an empty object by clicking the Add button. Further activation of this button results in new objects being added. With Object-No the user then chooses the object whose setting he wants to alter. If this object has a name this is entered under *Title*. Later, this title can be positioned anywhere in the drawing area. With *Mode* we can determine the type of the object to be surface. Other possibilities are: 3d curve (Curve), Contour diagram (Contour) and List for a list of points.

Depending on the chosen type, we can determine the plot-style (*Style*). For a surface we have the choice between Points (only the calculated sample points are drawn), WireFrame (wireframe model), HiddenLine (hidden lines are not shown), ColorPatches (like HiddenLine, only the surfaces (the so called patches) which constitute the graphic, are colored) and Transparent (like ColorPatches, only the patches are filled with a pattern. By using different patterns for different objects a feeling of transparency can be created). Dependent on the chosen style under *Mesh* the user can choose if the parameter lines are to be drawn in one direction only or in both directions. For our sphere we choose a hidden line model with color patches and parameter lines in both directions.

With the buttons Color, starting at and ending at we can choose the color of the object. For our model we only have to set the color of the mesh and for the patches.

The other settings are all concerned with the parameters of the object. For our choice (the sphere) we have to give three functions x, y and z each dependent on two variables u and v. These determine the x-, y- and z-coordinates of the object. For our sphere we choose spherical coordinates and define the parametrization as follows:

```
x(u,v)  := sin(u) * cos(v)
y(u,v)  := sin(u) * sin(v)
z(u,v)  := cos(u)
```

We set the range for u and v under *Range* as $0 \leq u \leq PI$ and $-PI \leq v \leq PI$, and with *Grid* we determine the number of sample points for both variables. To get a first impression, we choose 10 in both directions.

At first we do not change anything in the scene specific setttings, instead we press the Plot-Button to draw the graphic. The result is:

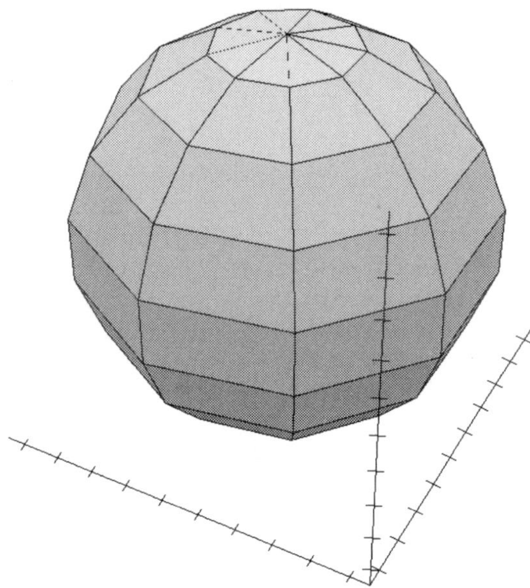

Figure 5.7: Sphere, first attempt

However, this does not look like a sphere. The problem is the small number of sample points. We could increase them with *Grid*, but this results in additional parameter lines being drawn. An alternative to this is to increase the *Smoothness* factor. With this, additional sample points can be stipulated that lead to interpolation between two visible plot points. We set the smoothness to 2 in both directions, press plot again and are rewarded with a much smoother representation of a sphere:

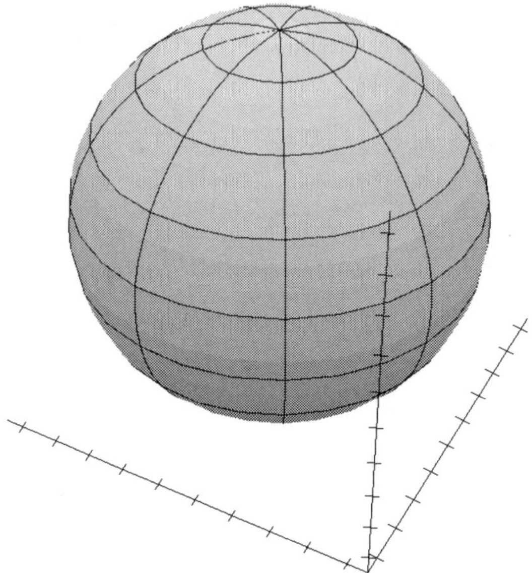

Figure 5.8: Smoothness = [2,2]

Now we should consider the scene specific characteristics. As with individual objects we can also give each scene a name. If coordinate axes are wanted these are chosen under *Axes*. With *Perspective* we can, as might be expected, change the perspective. This is determined by two points: CameraPoint and FocalPoint. The CameraPoint determines where the camera stands, FocalPoint gives the point to which the camera is directed. The user can choose between an automatic or manual perspective. The latter opens the perspective window in which the user can explicitly give both points, or the perspective can be altered with convenient sliders. The scaling routine to be used is chosen with the help of *Scaling*. With *Zooming* the visible part of the graphic can be changed. If manual is chosen then the graphic can be enlarged or scaled down step by step in a new window (Zooming Window). Further possible settings are grouped under *properties*. Among these are the axes legends and fonts for scene and object titles.

Chapter 6

List of functionality items provided by the system

The following is a short list of functions provided by the system at present. A detailed description of their functionality and syntax is found, together with various examples, in the reference manual.

Standard Library

Arithmetic functions	Functions for polynomials
Combinatoric functions	Generation of data structures
Manipulation of expressions	Evaluation of expressions
In- and output, graphics	Manipulation of strings
Functions with respect to procedures	Statistics
Debugging	Parallelism
Varia	

Arithmetic functions

`abs`	–	absolute value
`ceil`	–	rounding from above
`D`	–	differential operator
`diff`	–	differentiates an expression or polynomial
`expand`	–	expansion of an expression
`fact`	–	factorial
`float`	–	evaluation to a floating point number
`floor`	–	rounding from below
`frac`	–	fractional part of a number
`DIGITS`	–	number of digits for floating point numbers
`id`	–	identical function
`ifactor`	–	factorization of an integer
`igcd`	–	greatest common divisor
`igcdex`	–	extended Euclidean algorithm for integers
`ilcm`	–	least commom multiple
`isprime`	–	primality test
`ithprime`	–	delivers the i-th prime
`max`	–	maximum
`min`	–	minimum
`modp, mods`	–	functions for modular arithmetic
`nextprime`	–	delivers the next prime following its argument
`O`	–	order term
`phi`	–	Euler's ϕ-function
`powermod`	–	modular power
`random`	–	generation of a random number
`round`	–	rounding of a number
`sign`	–	sign of a number
`taylor`	–	Taylor series expansion
`trunc`	–	integer part of a number

Functions for polynomials

`coeff`	–	delivers the coefficients of a polynomial
`degree`	–	degree of a polynomial
`degreevec`	–	delivers the exponent of the leading term of a polynomial
`divide`	–	division of two polynomials
`evalp`	–	evaluation of a polynomial
`expr`	–	conversion of a polynomial into an expression
`genpoly`	–	generates a polynomial by a p-adic expansion
`icontent`	–	content of a polynomial
`iszero`	–	recognizes the zero polynomial
`lcoeff`	–	returns the leading coefficient of a polynomial
`lmonomial`	–	returns the leading monomial of a polynomial
`lterm`	–	returns the leading term of a polynomial
`mapcoeffs`	–	applies a function to the coefficients of a polynomial
`multcoeffs`	–	multiplies the coefficients of a polynomial with a factor
`norm`	–	norm of a polynomial
`nterms`	–	number of terms of a polynomial
`nthcoeff`	–	the n-th coefficient of a polynomial
`nthmonomial`	–	the n-th monomial of a polynomial
`nthterm`	–	the n-th term of a polynomial
`pdivide`	–	pseudo division of two polynomials
`poly`	–	generation of a polynomial
`Poly`	–	yields the domain of polynomials
`tcoeff`	–	the lowest coefficient of a polynomial

Combinatoric functions

`permute`	–	permutations of a list

Generation of data structures

`array`	–	definition of an array
`built_in`	–	definition of objects of type DOM_EXEC

`domain`	–	generation of a domain
`func_env`	–	definition of functional environments
`new`	–	generation of domain elements
`null`	–	delivers DOM_NULL
`poly`	–	generation of a polynomial
`table`	–	definition of a table

Manipulation of expressions

`anames`	–	returns the identifiers which are assigned
`append`	–	appends elements to a list
`cattype`	–	determines the data type
`contains`	–	tests existence of an element
`context`	–	evaluates in another context
`domattr`	–	access to entries of a domain
`domtype`	–	determines the data type
`extnops`	–	number of operands of a domain element
`extop`	–	returns the operands of a domain element
`extsubsop`	–	substitution of an operand in a domain element
`func`	–	considers an expression as function
`funcattr`	–	access to the third argument of a DOM_FUNC_ENV
`genident`	–	generates an unassigned identifier
`indets`	–	determines the unassigned identifiers contained in an expression
`index_val`	–	indexing without evaluation
`load`	–	returns those identifiers which have a value
`map`	–	applies a procedure to operands
`nops`	–	number of operands of an expression
`op`	–	returns the operands of an expression
`seq`	–	generates an expression sequence
`sort`	–	sorts the elements of a list
`subs`	–	substitution of complete subexpressions
`subsex`	–	extended substitution for arbitrary subexpressions
`subsop`	–	substitution of operands
`testtype`	–	checks the type of an expression

`type`	–	returns the type of an expression

Evaluation of expressions

`bool`	–	Boolean evaluation
`eval`	–	evaluation for some special functions
`evalassign`	–	assignment to an evaluation of the left side
`EVAL_STMT`	–	controls the evaluation of statements in expressions
`history`	–	exhibits the entries in the history table
`HISTORY`	–	length of history table
`hold`	–	prevents evaluation
`last`	–	access to last values
`level`	–	defines the substitution depth for evaluations
`LEVEL`	–	defines substitution depth for identifiers
`MAXLEVEL`	–	recognizes recursive definitions
`val`	–	value of an expression

In- and output, graphics

`fclose`	–	closes a file
`finput`	–	read-in of MuPAD-expressions from a file
`fopen`	–	opens a file
`fprint`	–	prints output to a file
`ftextinput`	–	line oriented read-in from a file
`input`	–	asks for interactive input of MuPAD-expressions
`loadlib`	–	loading of library procedures
`loadproc`	–	loads a procedure
`PATH`	–	search paths for files
`pathname`	–	returns a machine independent path name
`plot2d`	–	for graphical representation of 2-dimensional objects
`plot3d`	–	for graphical representation of 3-dimensional objects
`PRETTY_PRINT`	–	controls format of output
`print`	–	output to screen
`PRINTLEVEL`	–	printing depth for output of expressions and assignments

`protocol`	–	protocol of a MuPAD-session
`read`	–	read-in and execution of a file
`textinput`	–	asks for interactive input of text
`TEXTWIDTH`	–	width of a line
`write`	–	saves values of variables in a file
`Line Editor`	–	line editor for the terminal version

Manipulation of strings

`expr2text`	–	converts MuPAD-expressions to strings
`strlen`	–	length of a string
`strmatch`	–	comparison of two strings
`substring`	–	access to parts of a string
`tbl2text`	–	converts tables to strings
`text2tbl`	–	converts strings to tables
`text2expr`	–	converts strings to MuPAD-expressions
`text2list`	–	converts strings to lists

Functions with respect to procedures

`args`	–	access to the parameters of a procedure
`return`	–	exit from a procedure
`testargs`	–	controls the parameter check

Statistics

`bytes`	–	amount of memory used
`rtime`	–	measuring real time
`time`	–	measuring real CPU time
`stack`	–	stack

Debugging

`debug`	–	user controlled execution of procedures
`ERRORLEVEL`	–	controls the error check during evaluation
`error`	–	user defined error interrupt
`trace`	–	protocols the execution of procedures
`traperror`	–	intercepts errors

Parallelism

`global`	–	access to net variables
`readpipe`	–	reading from a pipe
`readqueue`	–	reading from a queue
`topology`	–	information about parallel structures
`writepipe`	–	writing into a pipe
`writequeue`	–	writing into a queue

Varia

`help`	–	access to a page of the help system
`reset`	–	resetting a MuPAD-session
`sysname`	–	delivers the name of the operating system
`system`	–	execution of a command from the operating system

Library Packages

Gröbner Basis
Orthogonal Polynomials

Gröbner Basis

`groebner`	–	Introduction for Gröbner basis package
`groebner::gbasis`	–	compute reduced Gröbner basis
`groebner::normalf`	–	reduced form modulo an ideal
`groebner::spoly`	–	compute the S polynomial of two polynomials

Orthogonal Polynomials

`orthpoly`	–	Introduction for the library package for orthogonal polynomials
`orthpoly::chebyshev1`	–	Chebyshev-polynomial of first kind
`orthpoly::chebyshev2`	–	Chebyshev-polynomial of second kind
`orthpoly::gegenbauer`	–	Gegenbauer-polynomial
`orthpoly::hermite`	–	Hermite-polynomial
`orthpoly::jacobi`	–	Jacobi-polynomial
`orthpoly::laguerre`	–	Laguerre-polynomial
`orthpoly::legendre`	–	Legendre-polynomial

Index

Progress in Theoretical Computer Science

Edited by
R.V. Book, University of California, Santa Barbara, CA, USA

Editorial Board:
E. Engeler, ETH Zemtrum, Zürich, Switzerland / **G. Huet,** INRIA, Le Chesnay, France
J.–P.Jouannaud, Université de Paris–Sud, Orsay, France / **R. Milner,** University of Edinburgh, Scotland
M. Nivat, Université de Paris VII, Paris, France / **M. Wirsing,** Universität Passau, Germany

Progress in Theoretical Computer Science is a series that focuses on the theoretical aspects of computer science and on the logical and mathematical foundations of computer science, as well as the applications of computer theory. It addresses itself to research workers and graduate students in computer and information science departments and research laboratories, as well as to departments of mathematics and electrical engineering where an interest in computer theory is found. The series publishes research monographs, graduate texts, and polished lectures from seminars and lecture series.